Multiple Regression
and the Analysis
of Variance
and Covariance

A Series of Books in Psychology

Editors: Jonathan Freedman
Gardner Lindzey
Richard F. Thompson

Multiple Regression and the Analysis of Variance and Covariance

---

Allen L. Edwards

UNIVERSITY OF WASHINGTON

W. H. Freeman and Company
New York

Library of Congress Cataloging in Publication Data

Edwards, Allen Louis.
Multiple regression and the analysis of variance
and covariance.

(A Series of books in psychology)
Bibliography: p.
Includes index.
1. Psychometrics. 2. Analysis of variance.
3. Regression analysis. I. Title. II. Series.
BF39.E32 1985 519.5′36′02415 84-25915
ISBN 0-7167-1703-4
ISBN 0-7167-1704-2 (pbk.)

Printed in the United States of America

4 5 6 7 8 9 VB 5 4 3 2 1 0 8 9 8 7

For Kyoko

Contents

Preface

At many universities and colleges, students in the behavioral sciences are required to take, in addition to a course in elementary statistical analysis, a second course concerned with the design of experiments and the analysis of experimental data. The standard texts for this second course, as offered in psychology departments, ordinarily provide excellent coverage of the analysis of variance but little on the multiple regression analysis of experimental data. I would not have the analysis of variance neglected, but I believe that a simultaneous consideration of both methods of data analysis adds considerably to the understanding of both the analysis of variance and multiple regression analysis.

In this book I have tried to show the correspondence between the analysis of variance and multiple regression analysis of experimental data. The book is intended for students who are currently taking or who have had a course in the analysis of variance and who wish to see the relationships between the analysis of variance and multiple regression analysis. Thus, the book might be used to supplement a standard text in a course on the analysis of variance or as a text in a course for students who have already been exposed to the analysis of variance.

I have tried to keep the notation in this book relatively simple so that it can be understood by the many students who get lost when confronted with multiple summations and multiple subscripts. This may offend some mathematicians and some statisticians, but it should please students who may have difficulty with elaborate and more precise notation.

The examples discussed in the text are simple and intentionally so. I am much more interested in having the reader understand the method of analysis than become involved in the calculations that real data sets would involve. It is for this reason that I have included examples in which the calculations for the most part can easily and quickly be accomplished with a hand calculator. I believe this to be most important for an understanding of the methods of analysis for those students with relatively little mathematical and statistical sophistication. The time for using the high-speed electronic computer is after one has an understanding of what the analysis is all about. Anyone who has been confronted by a student with a mass of computer output in hand asking for help in interpreting it must share this conviction. If a student has worked through various simple examples with a hand calculator, the problem of interpreting computer output often disappears. An instructor who wishes to do so can, of course, provide students with more realistic data sets. However, I am convinced that the time for this is after and not before the student is able to understand the analysis of relatively simple examples.

I do make certain demands of the reader. I expect the reader to have had an elementary course in statistics that included exposure to the t test for means and to the F test in the analysis of variance. It will also be helpful if the reader has already been exposed to linear regression and correlation involving two variables.

The reader of this book should also have a working knowledge of algebra. The reader who is deficient in algebraic skills will find it difficult to follow the various proofs and will have to rely on the equivalence of arithmetic calculations and to assume that this equivalence will hold for calculations involving different numbers. I have included a few proofs based on the calculus, but a knowledge of the calculus is not essential for an understanding of the material covered. I have also included some theorems based on matrix algebra without including the proofs. Again, a knowledge of matrix algebra is not essential for an understanding of the material covered.

I regard the exercises at the end of each chapter as an integral part of the book. I have included some proofs in these exercises. Many of the exercises were designed expressly to provide additional insight into the material covered in the chapter. Others were designed to test the reader's understanding of the material covered. The reader who wishes to gain the most from this book should not neglect the exercises at the end of each chapter.

The book is not intended as a general text on multiple regression analysis and does not cover many of the topics that would ordinarily be discussed in a general text. Nor is the book concerned with methods of multivariate analysis. As the title indicates, the book is concerned primarily with the relationship between multiple regression analysis and the analysis of variance and covariance.

Table I in the appendix is reprinted from Enrico. T. Federighi, Extended tables of the percentage points of student's t distribution, *Journal of the American Statistical Association*, 1959, **54**, 683–688, by permission of the American Statistical Association. Table II has been reprinted from George W. Snedecor and William G. Cochran, *Statis-*

tical Methods (6th ed.), copyright 1967 by Iowa State University Press, Ames, Iowa, by permission of the publisher.

For their careful reading of the manuscript of the first edition of this book, for checking the answers to the exercises at the end of the chapters, and for providing me with their reactions to and evaluations of the material contained in the first edition, I owe a special debt of gratitude to Jeffrey W. Anderson, Timothy Cahn, Felipe Gonzalez Castro, Sze-chak Chan, Virginia A. deWolf, William C. Follette, William Knowlton, Thomas J. Liu, Kyung-Hwan Min, Diane M. Morrison, Stevan Lars Nielsen, Jeanette Norris, Katherine P. Owen, Mark D. Pagel, Daniel L. Rock, Rachelle Dorn Rock, Margaret Rothman, Midori Yamagishi, and Captain Mitchell Zais.

Since the publication of the first edition, so many additional students have provided me with their reactions to and evaluations of the book that I cannot begin to list them all. I will say, however, that they are responsible for my rewriting various sections of the text in an attempt to achieve greater clarity of expression. In this edition I have also added one new chapter, reordered several chapters, and added some additional problems to the exercises.

Allen L. Edwards

July 1984

Multiple Regression and the Analysis of Variance and Covariance

1

The Equation of a Straight Line

1.1 Introduction

In many experiments in the behavioral sciences, performance of subjects on some dependent variable Y is studied as a function of some independent variable X. The values of the X, or independent, variable might consist of different dosages of a drug, different levels of reinforcement, different intensities of shock, or any other quantitative variable of experimental interest. The different values of the X variable *in an experiment* are ordinarily selected by the experimenter and are limited in number. They are also usually measured precisely and can be assumed to be without error. In general, we shall refer to the values of X in an experiment as fixed in that any conclusions based on the outcome of the experiment will be limited to the particular X values actually investigated.

For each of the X values, the experimenter obtains one or more observations of the dependent variable Y. In behavioral science experiments, the observations of the Y values are ordinarily associated with the performance of different subjects who have been assigned at random to the various values of the X variable.* For example, an experimenter might be interested in determining if there is any relationship between reaction time, the dependent variable Y, and amount of alcohol consumed, the independent variable X. For each dosage of alcohol or value of X included in the experiment, the experimenter will obtain reaction times from one or more subjects who have been randomly assigned to each dosage of alcohol. The objective of the experiment is to determine whether the Y values (or the average Y values, if more than one observation is obtained for each value of X) are related to the X values.

In this chapter we shall be concerned with two examples in which the Y values are linearly related to the X values.† By "linearly related" we mean that if the Y values are plotted against the X values, the resulting set of plotted points can be represented by a straight line. In the two examples considered, the relationship between Y and X is also "perfect," by which we mean that all of the plotted points fall exactly on a straight line. With a perfect linear relationship between Y and X, all of the Y values can be predicted exactly by an equation for a straight line.

The two examples are, however, not typical of experimental data in which the Y values rarely fall exactly on a straight line. But the examples will serve to introduce the equation for a straight line and the concepts of the slope and Y intercept of a straight line. A more realistic example will be given in the next chapter.

1.2 An Example of a Negative Relationship

Consider the values of X and Y given in Table 1.1 and the plot of the Y values against the X values shown in Figure 1.1. It is obvious that the points all fall on a straight line and that for each value of X the corresponding value of Y is equal to $-.3X$. We may express this rule in the following way:

$$Y = bX \tag{1.1}$$

where $b = -.3$ is a constant that multiplies each value of X. If each value of Y in Table 1.1 were exactly equal to the corresponding value of X, then the value of b would have to be equal to 1.0. If each

* An exception is a repeated-measure design in which the same subjects are tested with each value of X. This design is discussed in Chapter 10.

† Methods for studying nonlinear relationships are discussed in Chapter 8.

TABLE 1.1
$Y = -.3X$

X	Y
1	−.3
2	−.6
3	−.9
4	−1.2
5	−1.5
6	−1.8
7	−2.1
8	−2.4
9	−2.7
10	−3.0

value of Y were numerically equal to X but opposite in sign, then the value of b would have to be equal to -1.0. When the value of b is negative, the relationship between Y and X is also described as negative. When b is positive, the relationship between Y and X is also described as positive. With a negative relationship, the values of Y will decrease as the values of X increase, as shown in Figure 1.1.

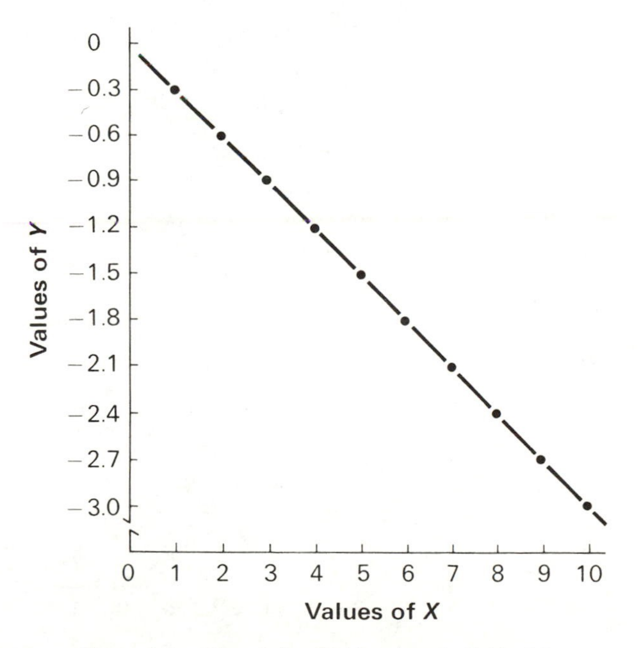

Figure 1.1 Plot of the (X, Y) values in Table 1.1.

1.3 An Example of a Positive Relationship

Now examine the values of Y and X in Table 1.2 and the plot of the Y values against the X values shown in Figure 1.2. It is obvious, in this instance, that the Y values increase as the X values increase and that the relationship between Y and X is positive. The rule or equation relating the Y values to the X values in Figure 1.2 has the general form

$$Y = a + bX \tag{1.2}$$

where b is again a constant that multiplies each value of X and a is a

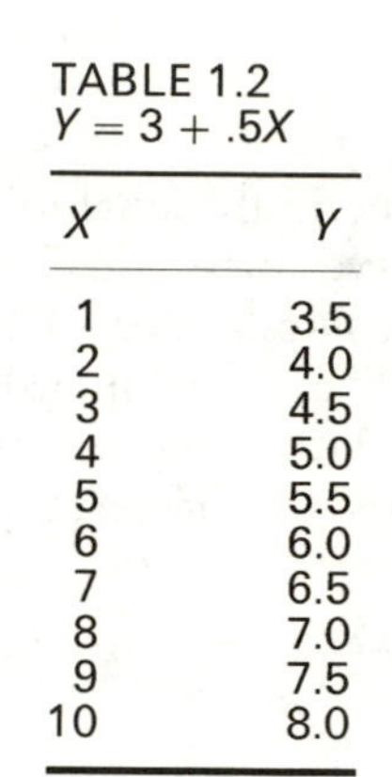

TABLE 1.2
$Y = 3 + .5X$

X	Y
1	3.5
2	4.0
3	4.5
4	5.0
5	5.5
6	6.0
7	6.5
8	7.0
9	7.5
10	8.0

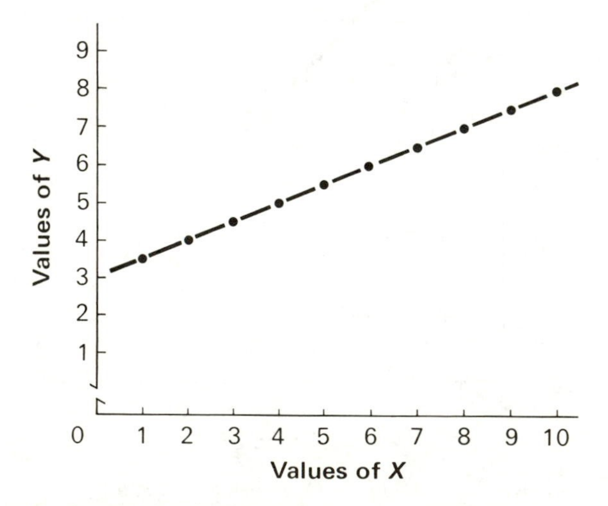

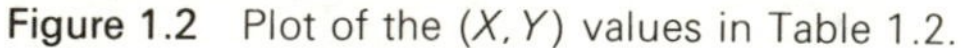

Figure 1.2 Plot of the (X, Y) values in Table 1.2.

constant that is added to each of the products.* For the values of Y and X given in Table 1.2, the value of b is equal to .5 and the value of a is equal to 3. Thus, when $X = 10$, we have $Y = 3 + (.5)(10) = 8$; when $X = 5$, we have $Y = 3 + (.5)(5) = 5.5$.

Both (1.1) and (1.2) are equations for a straight line. For example, we can take any arbitrary constants for a and b and any arbitrary set of X values. Then, substituting with the values of a and b and the values of X in (1.2), we can obtain a set of Y values. If these Y values are then plotted against the corresponding X values, the set of plotted points will fall on a straight line.

1.4 The Slope and Y Intercept

Note, in Figure 1.2, that as we move from 5 to 6 on the X scale, the corresponding increase in Y is from 5.5 to 6.0. An increase of 1 unit in X, in other words, results in .5 of a unit increase in Y. The constant b is simply the rate at which Y changes with unit change in X. The value of b can be obtained directly from Figure 1.2. For example, if we take any two points on the line with coordinates (X_1, Y_1) and (X_2, Y_2), then

$$b = \frac{Y_2 - Y_1}{X_2 - X_1} \tag{1.3}$$

Substituting in (1.3) with the coordinates (3, 4.5) and (7, 6.5), we have

$$b = \frac{6.5 - 4.5}{7 - 3} = .5$$

Similarly, for Figure 1.1, if we take the two points with coordinates (3, −.9) and (6, −1.8), we have

$$b = \frac{-1.8 - (-.9)}{6 - 3} = \frac{-.9}{3} = -.3$$

In geometry, (1.3) is known as a particular form of the equation of a straight line, and the value of b is called the slope of the straight line.

The nature of the additive constant a in (1.2) can readily be determined by setting X equal to zero. The value of a must then be the value of Y when X is equal to zero. If the straight line in Figure 1.2 were to be extended downward, we would see that the line would intersect the Y axis at the point $(0, a)$. The number a is called the Y intercept of the line. In Figure 1.2 it is easy to see that the value of a is equal

* There are other notations used for the equation of a straight line. Some examples are $Y = b_0 + b_1X_1$, $Y = bX + m$, and $Y = mx + b$. The additive constant is always the Y intercept and the multiplicative constant the slope of the line.

to 3. In Figure 1.1 it is also easy to see that when $X = 0$, then $Y = 0$, and, consequently, the value of a is also equal to zero. That is why the equation of the straight line in Figure 1.1 is equal to $Y = bX$ instead of $Y = a + bX$.

Exercises

1.1 In each of the following examples we give the coordinates (X, Y) of two points that fall on a straight line. For each example find the value of a, the Y intercept, and the value of b, the slope of the line.

(a) (1, 1.5), (4, 3) (b) (1, 3.0), (3, 2.5)
(c) (1, 3), (3, 1) (d) (2, 0), (4, 10)

1.2 Find the values of Y when X is equal to 1, 2, and 3 for each of these straight lines:

(a) $Y = 1 + 2X$ (b) $Y = 3 - .5X$
(c) $Y = .5 + X$ (d) $Y = -.8X$

1.3 If we have a straight line, and if (1, 3) and (2, 5) are two points on the line, find $(3, Y)$, $(4, Y)$, and $(5, Y)$.

1.4 We have the following points on a straight line: (2, 2), (1, 1), and (4, 4). What is the equation of the straight line?

1.5 The following points fall on a straight line: (20, 0), (16, 2), (10, 5), (6, 7), and (0, 10). What is the equation of the straight line?

1.6 The following points fall on a straight line: (1, −5.4), (2, −5.8), (3, −6.2), (4, −6.6), and (5, −7.0). What is the equation of the straight line?

1.7 Explain each of the following concepts:

dependent variable	positive relationship
independent variable	slope of a straight line
negative relationship	Y intercept

2

Linear Regression and Correlation

2.1 Introduction

When the values of Y are plotted against the corresponding values of X and all of the points fall precisely on a straight line, with nonzero slope equal to b, the relationship between the two variables is said to be perfect. This means that every observed value of Y will be given exactly by $Y = a + bX$. Although, as we pointed out in the preceding chapter, the values of the independent variable X may be assumed to be measured accurately and are fixed, this will, in general, not be true of the corresponding values of the dependent variable Y. When the values of Y obtained in an experiment are plotted against the corresponding values of X, the trend of the plotted points may appear to be linear, but the plotted points will ordinarily not fall precisely on any straight line that we might draw to represent the trend.

As a model for experimental data, we shall assume that the X values are fixed and measured without error but that the Y values are subject to random variation.* If the Y values do not fall precisely on any straight line, we shall select from the family of all possible straight lines the one that gives the "best fit" to the experimental data. This line is called the *regression line* of Y on X, and the equation of the line is called a *regression equation.* Because the various values of Y will not ordinarily all fall on the line of best fit or the regression line, we make a slight change in notation and write the corresponding regression equation as

$$Y' = a + bX$$

The value of b in the regression equation is called a *regression coefficient.*

2.2 The Mean and Variance of a Variable

Table 2.1 gives a set of ten paired (X, Y) values, and Figure 2.1 shows the plot of the Y values against the X values. It is apparent that the values of Y tend to increase as X increases, but it is also apparent that the values of Y will not fall on any straight line that we might draw to represent the trend of the points.

The horizontal line in Figure 2.1 corresponds to the mean of the Y values. The *mean* value of a variable is defined as the sum of the observed values divided by the number of values and is ordinarily represented by a capital letter with a bar over it. For the mean of the Y values we have

$$\bar{Y} = \frac{\Sigma Y}{n} \tag{2.1}$$

where ΣY indicates that we are to sum the n values of Y. For the mean of the Y values given in Table 2.1 we have

$$\bar{Y} = \frac{4 + 1 + 6 + \cdots + 15}{10} = \frac{71}{10} = 7.1$$

* Additional assumptions will be made about the Y values later.

TABLE 2.1 A set of ten paired (X, Y) values and the deviations, $Y - \bar{Y}$, and the squared deviations, $(Y - \bar{Y})^2$

	(1) X	(2) Y	(3) $Y - \bar{Y}$	(4) $(Y - \bar{Y})^2$
	1	4	−3.1	9.61
	2	1	−6.1	37.21
	3	6	−1.1	1.21
	4	2	−5.1	26.01
	5	5	−2.1	4.41
	6	11	3.9	15.21
	7	7	−0.1	.01
	8	11	3.9	15.21
	9	9	1.9	3.61
	10	15	7.9	62.41
Σ	55	71	0.0	174.90

Similarly, for the mean of the X values we have

$$\bar{X} = \frac{1 + 2 + 3 + \cdots + 10}{10} = \frac{55}{10} = 5.5$$

In Figure 2.1, vertical lines connect each plotted point and the mean (7.1) of the Y values. Each of these vertical lines represents a deviation of an observed value of Y from the mean of the Y values or the

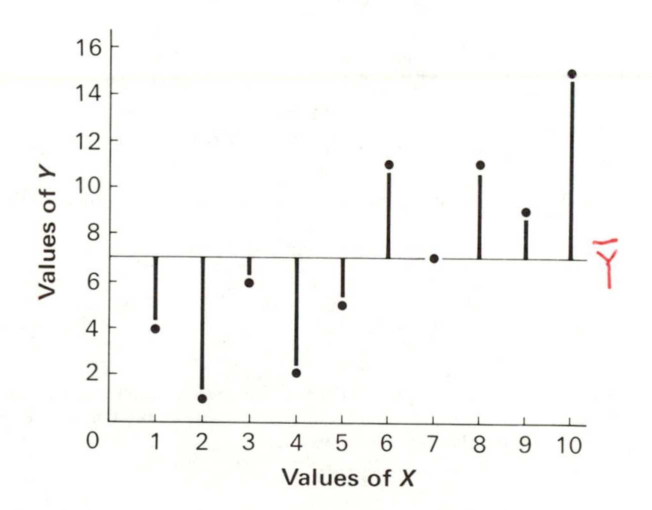

Figure 2.1 Plot of the (X, Y) values in Table 2.1, showing the deviations of the Y values from the mean of the Y values.

magnitude of $Y - \bar{Y}$. The values of $Y - \bar{Y}$ are given in Table 2.1, and we note that the algebraic sum of the deviations is equal to zero; that is,

$$\Sigma(Y - \bar{Y}) = 0$$

This will always be true regardless of the actual values of Y. For example, we will always have n values of $Y - \bar{Y}$, and, consequently, when we sum the n values we obtain

$$\Sigma(Y - \bar{Y}) = \Sigma Y - n\bar{Y}$$

because $\bar{Y}$ is a constant and will be subtracted n times. By definition, $\bar{Y} = \Sigma Y/n$, and, consequently, $n\bar{Y} = \Sigma Y$ and $\Sigma Y - n\bar{Y} = 0$.

The squares of the deviations, $(Y - \bar{Y})^2$, are shown in column 4 of Table 2.1. For the sum of the squared deviations we have

$$\Sigma(Y - \bar{Y})^2 = 174.90$$

It can be proved,* for any variable Y, that $\Sigma(Y - \bar{Y})^2$ will always be smaller than $\Sigma(Y - c)^2$, where c is any constant such that c is not equal to $\bar{Y}$. For example, if we subtract the constant $c = 7.0$ instead of $\bar{Y} = 7.1$ from each of the values of Y and sum the resulting squared deviations, we will find that $\Sigma(Y - 7.0)^2$ is larger than $\Sigma(Y - 7.1)^2$ because $c = 7.0$ is not equal to $\bar{Y} = 7.1$.

Dividing the sum of squared deviations from the mean by $n - 1$, we obtain a measure known as the *variance*. The variance is ordinarily represented by s^2, and for the Y variable we have

$$s^2 = \frac{\Sigma(Y - \bar{Y})^2}{n - 1} \tag{2.2}$$

or, for the values in Table 2.1,

$$s^2 = \frac{174.90}{10 - 1} = 19.4333$$

The square root of the variance is called the *standard deviation*. Thus,

$$s = \sqrt{\frac{\Sigma(Y - \bar{Y})^2}{n - 1}} \tag{2.3}$$

---

* A simple proof requires a knowledge of the rules of differentiation as taught in the first course of the calculus. We want to find the value of the constant c that minimizes $\Sigma(Y - c)^2$. Expanding this expression, we have

$$\Sigma Y^2 - 2c\Sigma Y + nc^2$$

Differentiating the expression with respect to c and setting the derivative equal to zero, we obtain $-2\Sigma Y + 2nc = 0$. Then we have $c = \bar{Y}$.

and for the Y values in Table 2.1 we have

$$s = \sqrt{\frac{174.90}{10 - 1}} = 4.408$$

Both the variance and the standard deviation measure the variability of the Y values about the mean of the Y values. When the variance and standard deviation are small, the values of Y will tend to have small deviations from the mean; when the variance and standard deviation are large, the values of Y will tend to have large deviations from the mean. If the values of a variable Y are subject to random variation then it will often be true that for a random sample of n values of Y we will find that approximately 95 percent of the values of Y will be included in the interval $\bar{Y} \pm 2s$. In our example it is obvious that all of the n values of Y fall within the interval $7.1 \pm (2)(4.408)$, or within the interval -1.716 to 15.916.

2.3 Finding the Values of *a* and *b* in the Regression Equation

Figure 2.2 is another plot of the Y values in Table 2.1 against the corresponding X values. This figure shows the horizontal line through the mean (7.1) of the Y values and also the vertical line through the mean (5.5) of the X values. These two lines will intersect or cross at the point with coordinates $(\bar{X}, \bar{Y})$, or $(5.5, 7.1)$. This point is shown in Figure 2.2 as a small open circle and is labeled B. Now suppose that we rotate the horizontal line represented by $\bar{Y}$ counterclockwise about the point

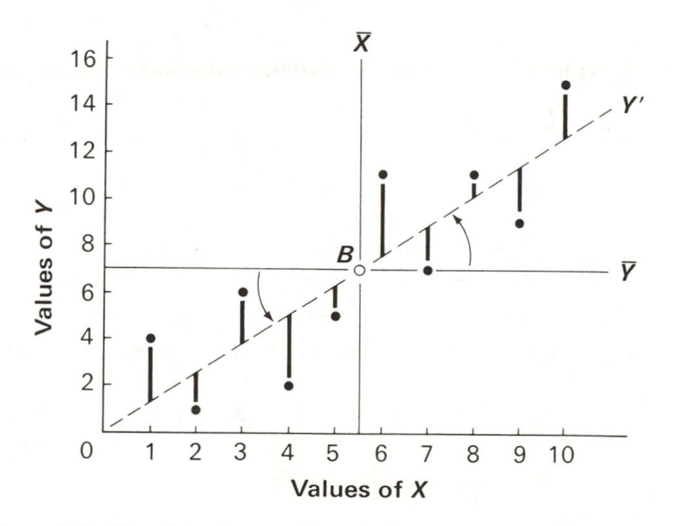

Figure 2.2 Plot of the (X, Y) values in Table 2.1 and the line of best fit.

B until we come to the dashed line Y'. This line, as drawn in Figure 2.2, is the regression line of Y on X, and the equation of this line, as we shall shortly see, is

$$Y' = a + bX \tag{2.4}$$

where

$$a = \bar{Y} - b\bar{X}$$

and

$$b = \frac{\Sigma(X - \bar{X})(Y - \bar{Y})}{\Sigma(X - \bar{X})^2}$$

Now, Y' as given by (2.4) can be equal to each and every observed value of Y if and only if there is a perfect linear relationship between Y and X, that is, only if the plotted points all fall precisely on a straight line. If a value of Y for a given value of X is not equal to Y', then we will have a residual

$$Y - Y' = Y - (a + bX) = e$$

The values of $e = Y - Y'$ are shown in Figure 2.2 by the vertical lines connecting the plotted points to the regression line Y', and the magnitudes of these residuals or deviations are given in column 7 of Table 2.2. Note that $\Sigma e = \Sigma(Y - Y') = 0$; we shall show why this is true later in this chapter.

The line drawn in Figure 2.2 was determined by the *method of least squares*. This criterion demands that the values of a and b in equation (2.4) be determined in such a way that the *residual sum of squares*

$$\Sigma(Y - Y')^2 = \Sigma[Y - (a + bX)]^2 = \Sigma e^2 \tag{2.5}$$

will be a minimum. It can be shown that the values of a and b in equation (2.4) that will make the residual sum of squares a minimum must satisfy the following two equations:*

$$an + b\Sigma X = \Sigma Y \tag{2.6}$$

---

* Again, a simple proof requires a knowledge of the rules of differentiation. By expanding the right side of (2.5), we obtain

$$\Sigma Y^2 - 2a\Sigma Y - 2b\Sigma XY + na^2 + 2ab\Sigma X + b^2\Sigma X^2$$

Differentiating this expression with respect to a and b and setting the derivatives equal to zero, we have

$$-2\Sigma Y + 2na + 2b\Sigma X = 0$$

and

$$-2\Sigma XY + 2a\Sigma X + 2b\Sigma X^2 = 0$$

from which we obtain (2.6) and (2.7).

and

$$a\Sigma X + b\Sigma X^2 = \Sigma XY \tag{2.7}$$

If we divide both sides of (2.6) by n and solve for a, we have

$$a = \bar{Y} - b\bar{X} \tag{2.8}$$

where $\bar{Y} = \Sigma Y/n$ is the mean of the Y values and $\bar{X} = \Sigma X/n$ is the mean of the X values. If we now substitute $\bar{Y} - b\bar{X}$ for a in (2.7), we have

$$\begin{aligned}\Sigma XY &= (\bar{Y} - b\bar{X})(\Sigma X) + b\Sigma X^2 \\ &= n\bar{X}\bar{Y} - bn\bar{X}^2 + b\Sigma X^2 \\ &= n\bar{X}\bar{Y} + b(\Sigma X^2 - n\bar{X}^2)\end{aligned}$$

Solving for b, we have

$$b = \frac{\Sigma XY - n\bar{X}\bar{Y}}{\Sigma X^2 - n\bar{X}^2}$$

or

$$b = \frac{\Sigma XY - (\Sigma X)(\Sigma Y)/n}{\Sigma X^2 - (\Sigma X)^2/n} \tag{2.9}$$

The necessary values for calculating a and b are given in Table 2.2. For the value of b we have

$$b = \frac{491 - (55)(71)/10}{385 - (55)^2/10} = \frac{100.5}{82.5} = 1.22$$

We also have $\bar{Y} = 71/10 = 7.1$ and $\bar{X} = 55/10 = 5.5$. Substituting with these values and with $b = 1.22$ in (2.8), we have

$$a = 7.1 - (1.22)(5.5) = .39$$

TABLE 2.2 Finding a line of best fit for the data in Table 2.1

	(1) X	(2) Y	(3) X^2	(4) Y^2	(5) XY	(6) Y'	(7) $Y - Y'$	(8) $(Y - Y')^2$
	1	4	1	16	4	1.61	2.39	5.7121
	2	1	4	1	2	2.83	−1.83	3.3489
	3	6	9	36	18	4.05	1.95	3.8025
	4	2	16	4	8	5.27	−3.27	10.6929
	5	5	25	25	25	6.49	−1.49	2.2201
	6	11	36	121	66	7.71	3.29	10.8241
	7	7	49	49	49	8.93	−1.93	3.7249
	8	11	64	121	88	10.15	0.85	0.7225
	9	9	81	81	81	11.37	−2.37	5.6169
	10	15	100	225	150	12.59	2.41	5.8081
Σ	55	71	385	679	491	71.00	0.00	52.4730

The regression equation for the data in Table 2.2 will then be

$$Y' = .39 + 1.22X$$

Note now that if we predict a value of Y corresponding to the mean of the X values, we obtain

$$Y' = .39 + (1.22)(5.5) = 7.1$$

which is equal to the mean of the Y values. The regression line will, therefore, pass through the point established by the mean of the X and Y values, or the point with coordinates $(\bar{X}, \bar{Y})$. This will be true for any regression line fitted by the method of least squares.

The predicted value of Y when X is equal to 2 will be

$$Y' = .39 + (1.22)(2) = 2.83$$

and when X is equal to 9 we have

$$Y' = .39 + (1.22)(9) = 11.37$$

The regression line will therefore pass through the points with coordinates (5.5, 7.1), (2, 2.83), and (9, 11.37). If we draw a line through these three points, it will be the regression line of Y on X, as shown in Figure 2.2. The values of Y' for each of the other values of X are given in column 6 of Table 2.2; these points will also fall on the regression line shown in Figure 2.2.

2.4 The Covariance and Orthogonal Variables

The numerator for the regression coefficient, as defined by (2.9), can be shown to be equal to the sum of the products of the deviations of the paired X and Y values from their respective means. Thus, we have

$$\begin{aligned}\Sigma(X - \bar{X})(Y - \bar{Y}) &= \Sigma XY - \bar{Y}\Sigma X - \bar{X}\Sigma Y + n\bar{X}\bar{Y} \\ &= \Sigma XY - n\bar{X}\bar{Y} - n\bar{X}\bar{Y} + n\bar{X}\bar{Y} \\ &= \Sigma XY - n\bar{X}\bar{Y}\end{aligned}$$

or

$$\Sigma(X - \bar{X})(Y - \bar{Y}) = \Sigma XY - \frac{(\Sigma X)(\Sigma Y)}{n} \tag{2.10}$$

It will be convenient to let $x = X - \bar{X}$ and $y = Y - \bar{Y}$ so that we also have

$$\Sigma xy = \Sigma(X - \bar{X})(Y - \bar{Y})$$

If we divide the sum of the products of the deviations of the paired X and Y values from their respective means by $n - 1$, the resulting measure is called the *covariance*. Thus, we have

$$c_{XY} = \frac{\Sigma(X - \bar{X})(Y - \bar{Y})}{n - 1} = \frac{\Sigma xy}{n - 1} \tag{2.11}$$

If the product sum, $\Sigma(X - \bar{X})(Y - \bar{Y})$, is equal to zero, then it is obvious that covariance and the regression coefficient will also be equal to zero. We shall refer to any two variables as *orthogonal* if their product sum or, equivalently, their covariance is equal to zero. Thus, if

$$\Sigma xy = \Sigma(X - \bar{X})(Y - \bar{Y}) = 0$$

X and Y are orthogonal variables. Similarly, if X_i and X_j are two X variables, then they will be orthogonal X variables provided that

$$\Sigma x_i x_j = \Sigma(X_i - \bar{X}_i)(X_j - \bar{X}_j) = 0$$

The denominator of the regression coefficient as defined by (2.9) can be shown to be equal to the sum of the squared deviations of the X values from the mean of the X values. Thus, we have

$$\begin{aligned} \Sigma(X - \bar{X})^2 &= \Sigma X^2 - 2\bar{X}\Sigma X + n\bar{X}^2 \\ &= \Sigma X^2 - \frac{(\Sigma X)^2}{n} \end{aligned} \tag{2.12}$$

The sum of squared deviations from the mean divided by $n - 1$ is, as we pointed out earlier, called the variance and is represented by s^2. Therefore, the variance of the X values is

$$s_X^2 = \frac{\Sigma(X - \bar{X})^2}{n - 1} = \frac{\Sigma x^2}{n - 1}$$

Then the regression coefficient as defined by (2.9) can also be written as

$$b = \frac{c_{XY}}{s_X^2} = \frac{\Sigma xy/(n - 1)}{\Sigma x^2/(n - 1)} = \frac{\Sigma xy}{\Sigma x^2} = \frac{\Sigma(X - \bar{X})(Y - Y\bar{Y})}{\Sigma(X - \bar{X})^2}$$

2.5 The Residual and Regression Sums of Squares

Column 8 in Table 2.2 shows the values of $e^2 = (Y - Y')^2$. For the sum of these squared values we have $\Sigma e^2 = \Sigma(Y - Y')^2 = 52.4730$; this sum of squared deviations is called the *residual sum of squares*. Let us see if we can gain some additional insight into the nature of this sum of squares. By definition,

$$Y' = a + bX$$

and because $a = \bar{Y} - b\bar{X}$, we have

$$Y' = \bar{Y} - b\bar{X} + bX$$

Summing over the n values, we have

$$\Sigma Y' = n\bar{Y} - nb\bar{X} + b\Sigma X$$

Because $nb\bar{X} = b\Sigma X$, the last two terms on the right side cancel, and we have

$$\Sigma Y' = n\bar{Y} = \Sigma Y$$

The sum of the predicted values, $\Sigma Y'$, is thus equal to the sum of the observed values, ΣY, and the mean of the predicted values, $\bar{Y}'$, must be equal to the mean of the observed values, $\bar{Y}$. We see that this is true for the data in Table 2.2, where $\Sigma Y = 71.0$ and $\Sigma Y' = 71.0$. Then it follows that

$$\Sigma e = \Sigma(Y - Y') = \Sigma Y - \Sigma Y' = 0$$

because we have just shown that $\Sigma Y = \Sigma Y'$.

We have shown that $Y' = a + bX = \bar{Y} - b\bar{X} + bX$. Then we also have

$$\begin{aligned} Y' &= \bar{Y} + bX - b\bar{X} \\ &= \bar{Y} + b(X - \bar{X}) \end{aligned}$$

and

$$Y' - \bar{Y} = b(X - \bar{X})$$

or

$$Y' - \bar{Y} = bx \tag{2.13}$$

and we know that $\Sigma(Y' - \bar{Y}) = b\Sigma x = 0$.

We also have

$$Y - Y' = (Y - \bar{Y}) - b(X - \bar{X})$$

or

$$Y - Y' = y - bx \tag{2.14}$$

and we also know that $\Sigma(Y - Y') = \Sigma y - b\Sigma x = 0$.

Then the deviation of Y from the mean of the Y values can be expressed as a sum of the two components given by (2.13) and (2.14), or as

$$Y - \bar{Y} = (Y' - \bar{Y}) + (Y - Y') \tag{2.15a}$$

or as

$$Y - \bar{Y} = bx + (y - bx) \tag{2.15b}$$

Squaring and summing over the n values of $Y - \bar{Y}$, we have

$$\Sigma(Y - \bar{Y})^2 = b^2\Sigma x^2 + \Sigma(y - bx)^2 + 2(b\Sigma xy - b^2\Sigma x^2)$$

Substituting an identity, $\Sigma xy/\Sigma x^2$, for b in the last term on the right, we see that

$$2\left[\frac{\Sigma xy}{\Sigma x^2}(\Sigma xy) - \frac{(\Sigma xy)^2}{(\Sigma x^2)^2}(\Sigma x^2)\right] = 0$$

Consequently, we have

$$\Sigma(Y - \bar{Y})^2 = \Sigma(Y' - \bar{Y})^2 + \Sigma(Y - Y')^2 \qquad (2.16a)$$

or

$$\Sigma(Y - \bar{Y})^2 = b^2\Sigma x^2 + \Sigma(y - bx)^2 \qquad (2.16b)$$

For the first term on the right in (2.16b) we note that

$$\Sigma(Y' - \bar{Y})^2 = b^2\Sigma x^2 = \frac{(\Sigma xy)^2}{(\Sigma x^2)^2}(\Sigma x^2) = \frac{(\Sigma xy)^2}{\Sigma x^2} \qquad (2.17)$$

This sum of squares is called the *regression sum of squares.* In our example we have $\Sigma xy = 100.5$ and $\Sigma x^2 = 82.5$. Then for the regression sum of squares we have

$$\Sigma(Y' - \bar{Y})^2 = \frac{(100.5)^2}{82.5} = 122.427$$

rounded.

For the second term on the right in (2.16b) we note that

$$\begin{aligned}\Sigma(Y - Y')^2 &= \Sigma(y - bx)^2 \\ &= \Sigma y^2 + b^2\Sigma x^2 - 2b\Sigma xy \\ &= \Sigma y^2 - \frac{(\Sigma xy)^2}{\Sigma x^2}\end{aligned} \qquad (2.18)$$

This sum of squares is called the *residual sum of squares.* In our example, we have

$$\Sigma(Y - Y')^2 = 174.900 - \frac{(100.5)^2}{82.5} = 52.473$$

rounded.

We see, therefore, that if there is a linear relationship between Y and X, the total sum of squared deviations of the Y values from the mean of the Y values can be partitioned into two orthogonal components, the sum of squares for linear regression, $\Sigma(Y' - \bar{Y})^2 = (\Sigma xy)^2/\Sigma x^2$, and the residual sum of squares, $\Sigma(Y - Y')^2 = \Sigma y^2 - (\Sigma xy)^2/\Sigma x^2$. It will be convenient to designate these three sums of squares as SS_{tot}, SS_{reg}, and SS_{res}, respectively. Then the following three equations are identical:

$$\Sigma(Y - \bar{Y})^2 = \Sigma(Y' - \bar{Y})^2 + \Sigma(Y - Y')^2 \tag{2.19a}$$

$$\Sigma y^2 = \frac{(\Sigma xy)^2}{\Sigma x^2} + \left[\Sigma y^2 - \frac{(\Sigma xy)^2}{\Sigma x^2}\right] \tag{2.19b}$$

$$SS_{tot} = SS_{reg} + SS_{res} \tag{2.19c}$$

If there is a perfect linear relationship between Y and X so that all of the plotted points fall precisely on a straight line, then SS_{res} will be equal to zero and SS_{reg} will be equal to SS_{tot}.

If there is no tendency for the Y values to be linearly related to the X values, then the regression coefficient b and SS_{reg} will be equal to zero, and SS_{res} will be equal to SS_{tot}. An important index of the degree to which the Y and X values are linearly related is obtained by dividing both sides of any one of the above three equations by $SS_{tot} = \Sigma y^2$. Then we see that

$$1 = \frac{(\Sigma xy)^2}{(\Sigma x^2)(\Sigma y^2)} + \left[1 - \frac{(\Sigma xy)^2}{(\Sigma x^2)(\Sigma y^2)}\right] \tag{2.20}$$

and, as we show in the next section, (2.20) reduces to

$$1 = r^2 + (1 - r^2)$$

where r is the correlation coefficient between Y and X.

2.6 The Correlation Coefficient

The *correlation coefficient* between Y and X can be defined as

$$r = \frac{\Sigma xy}{\sqrt{(\Sigma x^2)(\Sigma y^2)}} \tag{2.21}$$

For the data in Table 2.2 we have

$$\Sigma xy = \Sigma XY - \frac{(\Sigma X)(\Sigma Y)}{n} = 491 - \frac{(55)(71)}{10} = 100.5$$

and

$$\Sigma x^2 = \Sigma X^2 - \frac{(\Sigma X)^2}{n} = 385 - \frac{(55)^2}{10} = 82.5$$

and

$$\Sigma y^2 = \Sigma Y^2 - \frac{(\Sigma Y)^2}{n} = 679 - \frac{(71)^2}{10} = 174.90$$

Then the square of the correlation coefficient will be

$$r^2 = \frac{(100.5)^2}{(82.5)(174.9)} = .70$$

rounded, and

$$r = \frac{100.5}{\sqrt{(82.5)(174.90)}} = .84$$

rounded.

As (2.20) indicates, the square of the correlation coefficient is simply the proportion of the total sum of squares that can be accounted for by the linear regression of Y on X, or

$$r^2 = \frac{SS_{reg}}{SS_{tot}} = \frac{(\Sigma xy)^2}{(\Sigma x^2)(\Sigma y^2)}$$

Similarly, $1 - r^2$ is the proportion of the total sum of squares that cannot be accounted for or that is independent of the linear regression of Y on X; that is,

$$1 - r^2 = \frac{SS_{res}}{SS_{tot}} = 1 - \frac{(\Sigma xy)^2}{(\Sigma x^2)(\Sigma y^2)}$$

For the paired (X, Y) values in Table 2.2, approximately 70 percent of the total sum of squares can be accounted for by the linear regression of Y on X; and approximately 30 percent of the total sum of squares is independent of the regression of Y on X.

2.7 Mean Squares

It is a fundamental theorem of linear regression that SS_{tot} can always be partitioned into the two orthogonal components, SS_{reg} and SS_{res}. The two components are orthogonal in the sense described earlier, that is, because the product sum

$$\Sigma(Y' - \bar{Y})(Y - Y') = 0$$

Associated with each of the three sums of squares are divisors, called *degrees of freedom*, abbreviated d.f. When a sum of squares is divided by its degrees of freedom, the resulting value is called a *mean square*. We know that the divisor for SS_{tot} is $n - 1$, and the total sum of squares is said to have $n - 1$ degrees of freedom. Then we have the identity

$$s_Y^2 = MS_{tot} = \frac{\Sigma(Y - \bar{Y})^2}{n - 1} \tag{2.22}$$

Now consider what would happen if we had only two pairs of (X, Y) values. In this cases, SS_{tot} would have $n - 1 = 1$ degree of freedom.

With only two plotted points, both would fall on a straight line, and SS_{reg} would be equal to SS_{tot} and SS_{res} would be equal to zero and would have $n - 2 = 0$ degrees of freedom. It must be true, then, that SS_{reg} has 1 degree of freedom. Then the mean square for regression will be

$$MS_{reg} = SS_{reg}/1 \tag{2.23}$$

and the residual mean square will be*

$$MS_{res} = \frac{SS_{res}}{n - 2} \tag{2.24}$$

When a total sum of squares is partitioned into a set of orthogonal components, the orthogonal components are additive; that is, they will add up to the total sum of squares. When sums of squares are orthogonal, the degrees of freedom associated with each of the sums of squares are also additive; that is, they will add up to the total degrees of freedom. Note that it is sums of squares and degrees of freedom that are additive, not mean squares. Thus, we have

$$SS_{tot} = SS_{reg} + SS_{res}$$

and

$$n - 1 = 1 + (n - 2)$$

but

$$MS_{tot} \neq MS_{reg} + MS_{res}$$

2.8 The Nature of r^2 and $1 - r^2$

Because $r^2 = (\Sigma xy)^2/(\Sigma x^2)(\Sigma y^2)$, we can, as we know, rewrite (2.20) as

$$1 = r^2 + (1 - r^2) \tag{2.25}$$

We also know that

$$SS_{tot} = SS_{reg} + SS_{res}$$

and that

$$r^2 = \frac{SS_{reg}}{SS_{tot}}$$

---

* In later chapters we shall have linear equations involving more than one X variable. The degrees of freedom for SS_{reg} will be equal to k, where k is the number of X variables in the regression equation, and the degrees of freedom for SS_{res} will be equal to $n - k - 1$, where n is the total number of observations. In this chapter we have only one X variable, and $k = 1$. Thus, we have $n - k - 1 = n - 2$. For a further discussion of degrees of freedom, see Exercise 2.21.

Because SS_{reg} can be equal to but never larger than SS_{tot}, it is obvious that r^2 can never be greater than 1 and, consequently, r can range only within the limits -1.00 to 1.00. When r is equal to -1.00 or 1.00, all of the plotted points will fall on the regression line, and SS_{reg} will be equal to SS_{tot}. Figure 2.3(a) and (b) are plots where $r = 1.00$ and $r = -1.00$, respectively. In Figure 2.3(c), r is equal to .86, and in Figure 2.3(d), r is equal to $-.34$. As we might guess, the value of r^2 provides an index of the degree to which a set of plotted points clusters about the regression line. The closer the points fall along the regression line, the larger the value of r^2 and the greater the proportion of the total sum of squares accounted for by the linear regression of Y on X. When the value of r^2 is small, as in Figure 2.3(d), the plotted points will show considerable scatter about the regression line, and the proportion of the total sum of squares accounted for by the linear regression of Y on X will be small. For the data plotted in Figure 2.3(d) we have $r^2 = SS_{reg}/SS_{tot} = (-.34)^2 = .1156$.

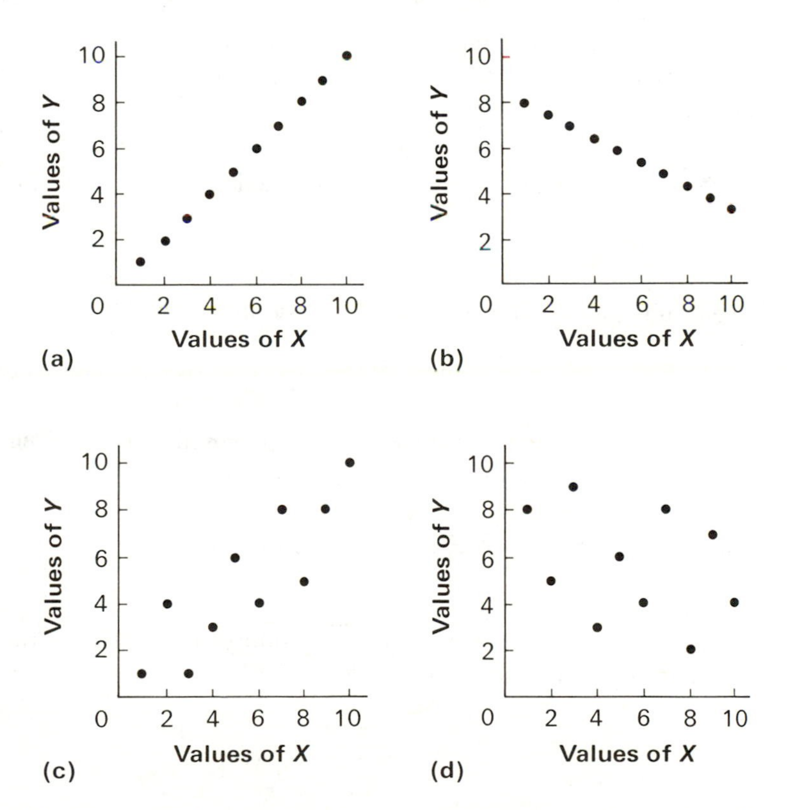

Figure 2.3 Plots of (X, Y) values for which (a) $r = 1.00$, (b) $r = -1.00$, (c) $r = .86$, and (d) $r = -.34$.

If we multiply both sides of (2.25) by s_Y^2, we obtain

$$s_Y^2 = r^2 s_Y^2 + (1 - r^2)s_Y^2 \tag{2.26}$$

Similarly, if we multiply both sides of (2.25) by s_X^2, we have

$$s_X^2 = r^2 s_X^2 + (1 - r^2)s_X^2 \tag{2.27}$$

and r^2 is the proportion of variance that X and Y have in common. If X and Y are orthogonal variables, then r will be equal to zero and X and Y will have no variance in common.

2.9 The Correlation Coefficient and the Regression Coefficient

Although (2.21) can be used as a definition of the correlation coefficient, it is more often defined as the covariance between X and Y divided by the product of the standard deviations of X and Y, or as

$$r = \frac{c_{XY}}{s_X s_Y} \tag{2.28}$$

If we multiply both the numerator and denominator of (2.28) by $(n - 1)$, we obtain r as defined by (2.21). From (2.28) we also have the important identity

$$c_{XY} = r s_X s_Y \tag{2.29}$$

We showed earlier that the regression coefficient could be written as

$$b = \frac{c_{XY}}{s_X^2}$$

Then, substituting an identity for the covariance, we obtain

$$b = \frac{r s_X s_Y}{s_X^2} = r\frac{s_Y}{s_X} \tag{2.30}$$

The regression coefficient, as (2.30) shows, can be equal to the correlation coefficient if and only if $s_Y = s_X$.

2.10 Standard Error of the Regression Coefficient

We assume that the values of X in a regression problem are fixed constants and that the values of Y are subject to random variation. Without any loss of generality we let $x = X - \bar{X}$, and we note that

$$\begin{aligned}\Sigma(X - \bar{X})Y &= \Sigma XY - \bar{X}\Sigma Y \\ &= \Sigma XY - \frac{(\Sigma X)(\Sigma Y)}{n} \\ &= \Sigma xy\end{aligned}$$

Then, with $x = (X - \bar{X})$, the regression coefficient can be written as

$$b = \frac{\Sigma xY}{\Sigma x^2} = \frac{1}{\Sigma x^2}(x_1 Y_1 + x_2 Y_2 + \cdots + x_i Y_i + \cdots + x_n Y_n) \quad (2.31)$$

and b is just a weighted linear sum of the Y's. If the values of x_i are fixed constants, then $k = \Sigma x^2$ will also be a fixed constant. The values of Y_i associated with a fixed value of x_i in repeated random samples will have a population variance, $\sigma^2_{Y \cdot X}$, that is assumed to be the same for each fixed value of x_i. Then the variance of any one of the terms in (2.31) will be

$$\frac{x_i^2 \sigma^2_{Y \cdot X}}{k^2}$$

because multiplying a random variable by a constant multiplies the variance by the square of the constant, and dividing a random variable by a constant divides the variance by the square of the constant. The Y's associated with the different values of X are assumed to be independent random variables, and the variance of a sum of independent random variables is simply the sum of their variances. Then the variance of b will be equal to

$$\sigma_b^2 = \frac{\Sigma x^2 \sigma^2_{Y \cdot X}}{k^2} = \frac{\Sigma x^2 \sigma^2_{Y \cdot X}}{(\Sigma x^2)^2} = \frac{\sigma^2_{Y \cdot X}}{\Sigma x^2}$$

The sample estimate of $\sigma^2_{Y \cdot X}$ is MS_{res}. Consequently, the sample estimate of the variance of b will be

$$s_b^2 = \frac{MS_{res}}{\Sigma x^2}$$

and the standard error of b will be

$$s_b = \sqrt{\frac{MS_{res}}{\Sigma x^2}} \quad (2.32)$$

2.11 Tests of Significance

The population value of the regression coefficient corresponding to the sample value b is represented by β. Under the null hypothesis that in the population the Y values are not linearly related to the X values, we have $\beta = 0$. The test of this null hypothesis that is ordinarily reported in textbooks is the t test defined by

$$t = \frac{b}{\sqrt{MS_{res}/\Sigma x^2}} \tag{2.33}$$

where $s_b = \sqrt{MS_{res}/\Sigma x^2}$ is the standard error of b. This t test will have 1 degree of freedom for the numerator and $n - k - 1$, where k is equal to 1, degrees of freedom for the denominator. Because t^2 with 1 and $n - 2$ degrees of freedom is equal to F with 1 and $n - 2$ degrees of freedom, we also have*

$$F = \frac{b^2}{MS_{res}/\Sigma x^2} \tag{2.34}$$

The population value of the correlation coefficient corresponding to the sample value r is represented by ρ. If the Y values in the population are not linearly related to the X values, that is, if $\beta = 0$, then it will also be true that $\rho = 0$. The test of the null hypothesis that $\rho = 0$ is ordinarily reported in textbooks as a t test defined by

$$t = \frac{r}{\sqrt{1 - r^2}} \sqrt{n - 2} \tag{2.35}$$

with 1 and $n - 2$ degrees of freedom. Again, because t^2 with 1 and $n - 2$ degrees of freedom is equivalent to F with 1 and $n - 2$ degrees of freedom, we also have

$$F = \frac{r^2}{1 - r^2}(n - 2) \tag{2.36}$$

We now show that the F tests of b and r are equal to each other and that both are equal to

$$F = \frac{MS_{reg}}{MS_{res}} \tag{2.37}$$

We have $b^2 = (\Sigma xy)^2/(\Sigma x^2)^2$, and, substituting in (2.34), we obtain

$$F = \frac{(\Sigma xy)^2/(\Sigma x^2)^2}{MS_{res}/\Sigma x^2} = \frac{(\Sigma xy)^2/\Sigma x^2}{MS_{res}} = \frac{MS_{reg}}{MS_{res}}$$

Similarly, we have $r^2 = (\Sigma xy)^2/(\Sigma x^2)(\Sigma y^2)$, and, substituting in (2.36), we obtain

$$F = \frac{(\Sigma xy)^2/(\Sigma x^2)(\Sigma y^2)}{1 - r^2}(n - 2)$$

* For any t test, there is, in general, a corresponding F test such that $F = t^2$.

But we also know that $SS_{res} = \Sigma y^2(1 - r^2)$, and, multiplying the numerator and denominator of the above expression by $\Sigma y^2/(n - 2)$, we have

$$F = \frac{(\Sigma xy)^2/\Sigma x^2}{\Sigma y^2(1 - r^2)/(n - 2)} = \frac{MS_{reg}}{MS_{res}}$$

In our example, we have $MS_{reg} = 122.427$ and $MS_{res} = 52.473/8 = 6.559$. Then

$$F = \frac{122.427}{6.559} = 18.665$$

which is a significant value with 1 and 8 d.f. and with $\alpha = .01$. We would obtain exactly the same value of F using (2.34) or (2.36). Consequently, we reject the null hypothesis that $\beta = 0$ and also the null hypothesis that $\rho = 0$.

The assumptions involved in the tests of significance of b and r and MS_{reg} are that the values of Y for each fixed value of X are independent of the values of Y for the other fixed values of X. For each fixed value of X, the Y's are assumed to be normally distributed with the same variance $\sigma^2_{Y \cdot X}$. The ordered values of $\mu_{Y \cdot X}$, the population means of the Y values for each of the X values, are assumed to fall on a straight line with slope equal to β. Then MS_{res} will be an unbiased estimate of $\sigma^2_{Y \cdot X}$.

2.12 Plotting of the Residuals, $Y - Y'$, Against X

A plot of the residuals, $e = Y - Y'$, against the values of X will often provide valuable and interesting information about the relationship between Y and X in addition to the usual statistics that we may calculate.* Figure 2.4 shows a plot of the residuals for the data in Table 2.2 against the values of X. The important thing to notice about the residuals is that they show no apparent trend or pattern with changes in X. This is the sort of graph we would expect to find if the relationship between Y and X is linear and if the various assumptions made in a regression analysis are met.

Figure 2.5(a) shows a plot of Y against X for $n = 6$ observations. If the regression equation, $Y' = a + bx$, is obtained for these data and the residuals, $e = Y - Y'$, are plotted against X, the trend of the

* It is often desirable to divide each of the residuals by $\sqrt{MS_{res}}$ before plotting them. Some computer programs do this automatically when plots of residuals are requested. For a more detailed discussion of graphical plots of residuals, see Draper and Smith (1966) and Anscombe (1973).

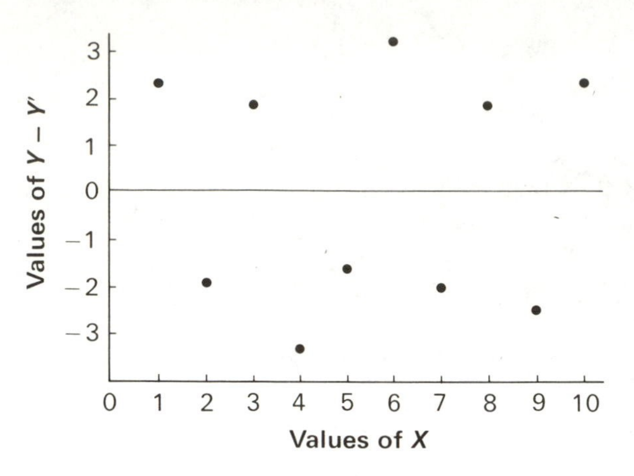

Figure 2.4 Plot of the residuals, $e = Y - Y'$, given in Table 2.2 against the values of X.

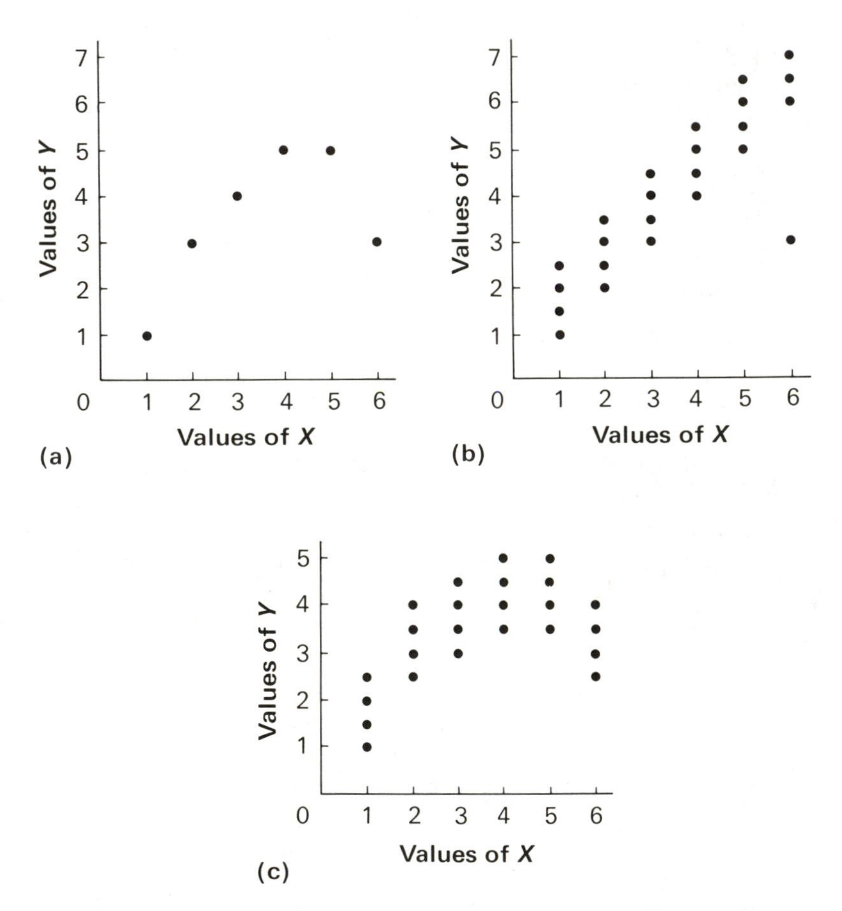

Figure 2.5 (a) Plot of $n = 6$ values of Y against X. (b) A possible outcome of the experiment in (a) with additional replications. (c) Another possible outcome of the experiment in (a) with additional replications.

residuals will be much the same as that shown in Figure 2.5(a). Now $n = 6$ is a small sample, and at least two possibilities exist: (1) Either the relationship between Y and X is not linear over the range of X values studied, or (2) the point with coordinates (6, 3) is an outlier. Suppose, however, that we had three additional observations of Y for each of the fixed values of X. Figure 2.5(b) shows one possible outcome of this experiment, and Figure 2.5(c) shows another. For the points plotted in Figure 2.5(b), it would appear that the relationship between Y and X is essentially linear and that the point with coordinates (6, 3) is an outlier. On the other hand, for the points plotted in Figure 2.5(c), it would appear that the relationship between Y and X is not linear, and the linear regression model, $Y' = a + bX$, is not realistic.

With reasonably large samples and with all of the assumptions of the regression analysis satisfied, a plot of the residuals against X should be similar to the plot shown in Figure 2.6(a). There is in this plot no systematic tendency for the residuals to change as X changes. In Figure

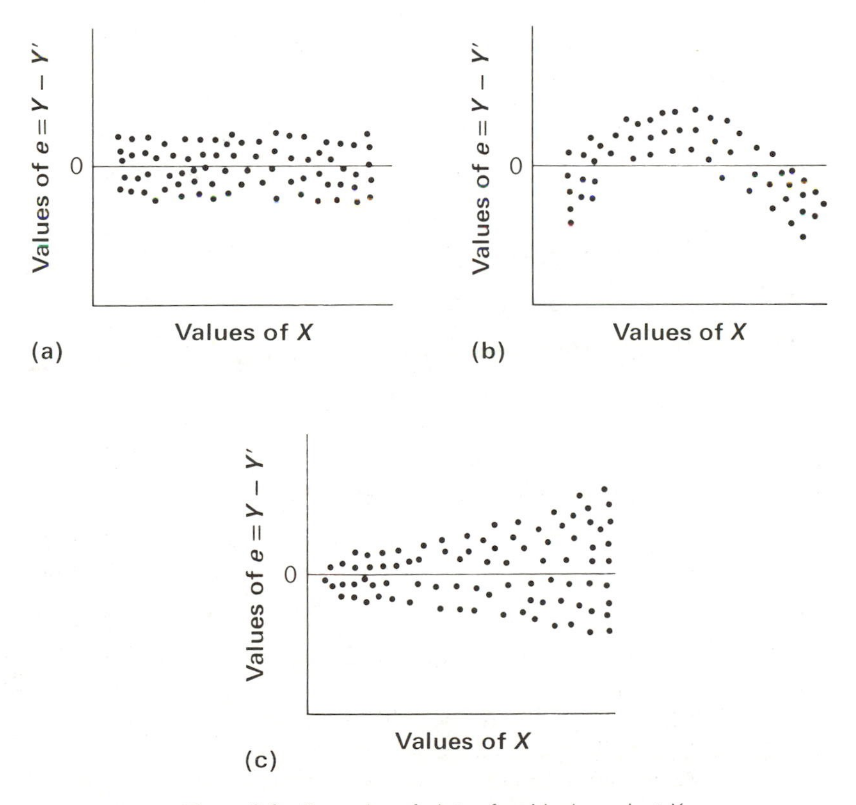

Figure 2.6 Examples of plots of residuals against X.

2.6(b), the plot of the residuals against X indicates that a linear regression model is not appropriate. Figure 2.6(c) illustrates the case where the variance of the residuals increases with increase in X. The regression analysis assumes that for each fixed value of X the population variance of $e = Y - Y'$ is equal to the same value $\sigma^2_{Y \cdot X}$. Thus, for the plot in Figure 2.6(c) we probably have a violation of the assumption of constant variance for Y for the various values of X.

Exercises

2.1 Make a plot of the following paired (X, Y) values: (1, 1), (2, 2), (3, 3), (4, 4), and (6, 1).

(a) What is the value of the correlation coefficient?

(b) What is the value of the correlation coefficient if the point with coordinates (6, 1) is omitted?

2.2 Make a plot of the following paired (X, Y) values: (1, 1), (1, 3), (2, 2), (3, 1), (3, 3), and (8, 8).

(a) What is the value of the correlation coefficient?

(b) What is the value of the correlation coefficient if the point with coordinates (8, 8) is omitted?

2.3 If $r = -1.00$, what is the value of SS_{reg}, and what is the value of SS_{res}?

2.4 Under what conditions will the regression coefficient be equal to the correlation coefficient?

2.5 Prove that the product sum $\Sigma(Y' - \bar{Y})(Y - Y')$ is equal to zero.

2.6 Prove that the product sum $\Sigma(X - \bar{X})(Y - Y')$ is equal to zero.

2.7 What is the approximate value of r if it is true that SS_{reg} is equal to SS_{res}?

2.8 If the correlation coefficient between X and Y is equal to .70, and if $Z = Y + 3$, what will be the value of the correlation coefficient between X and Z? See if you can prove that your answer is correct.

2.9 If the correlation coefficient between X and Y is .50, and if $Z = (Y - \bar{Y})/s_Y$, what will be the value of the correlation coefficient between X and Z?

2.10 We have the following paired (X, Y) values: (−2, 4), (−1, 1), (0, 0), (1, 1), and (2, 4).

(a) Either make a plot and guess the value of the correlation coefficient between X and Y, or calculate the value.

(b) Is Y functionally related to X? If so, what is the rule that relates Y to X?

2.11 In a class of 100 students, the average score on a midterm examination was 60; 50 students had scores below 60 and 50 had scores above 60. The mean scores on the midterm for these two groups were 50 and 70, respectively, a difference between the two means of 20 points. The correlation coefficient between scores on the midterm and scores on the final examination was .70. The mean score for all students on the final examination was also equal to 60. Both examinations had standard deviations equal to 10.0. Would you expect the difference between the means of the two groups on the final examination to be smaller than, larger than, or equal to the difference on the midterm examination? Explain why.

2.12 We have the following paired (X, Y) values: $(1, 1)$, $(2, 3)$, $(3, 4)$, $(4, 5)$, $(5, 5)$, and $(6, 3)$.

(a) Find the regression equation $Y' = a + bX$.

(b) Calculate the residuals, $e = Y - Y'$, and plot these against the values of X. What does the plot of residuals suggest? Use an adequate scale for the residuals, one that is approximately equal to the X scale.

2.13 If the regression coefficient is equal to zero, will the covariance also be equal to zero?

2.14 If the regression coefficient is equal to zero, will the correlation coefficient also be equal to zero?

2.15 Find the regression equation $Y' = a + bX$ for the paired (X, Y) values in the following table. Plot the residuals, $Y - Y'$, against the values of X. Do the residuals show any strong pattern?

X	Values of Y
2	3, 6
4	2, 4, 8
6	5, 7, 10
8	5, 8, 10
10	8, 12
12	5, 9, 11

2.16 For a set of paired (X, Y) values we have $SS_{tot} = 40$ and $r = .60$.

(a) What is the value of SS_{reg}?

(b) What is the value of SS_{res}?

(c) What proportion of the variance of X and the variance of Y is shared variance?

2.17 If X and Y are orthogonal variables, what do we know about the covariance of X and Y?

2.18 If two variables are orthogonal variables, what will the correlation coefficient between the two variables be equal to?

2.19 If the t test of the regression coefficient is significant with $\alpha = .05$, will $F = MS_{reg}/MS_{res}$ also be significant with $\alpha = .05$? Explain why or why not.

2.20 Prove that $\Sigma(X - \bar{X})Y$ is equal to $\Sigma(X - \bar{X})(Y - \bar{Y})$.

2.21 Suppose that we have n paired (X, Y) values.

(a) We will have n values of $Y - \bar{Y}$. If we know $n - 1$ of these values, will we also know the remaining value? Explain why or why not.

(b) We will also have n values of $Y' - \bar{Y}$. We assume that the values of X are known. The value of b is not known. If we know any one of the values of $Y' - \bar{Y}$, not equal to zero, then all of the remaining $n - 1$ values of $Y' - \bar{Y}$ are known. Explain why this is so.

2.22 Explain each of the following concepts:

correlation coefficient	regression sum of squares
covariance	regression line
mean	residual
mean square	residual sum of squares
orthogonal variables	standard deviation
regression coefficient	total sum of squares
regression equation	variance

3

Standardized Variables and Partial and Semipartial Correlation Coefficients

3.1 Introduction

We define a *standardized variable* as one that has a mean equal to zero and a variance and standard deviation equal to 1. As we shall show later in the chapter, the covariance between two standardized variables is equal to the correlation coefficient between the two variables. Because of these properties, standardized variables play an important role in linear regression and correlation. Algebraic proofs involving standardized variables are often much easier to develop and follow than the same proofs involving nonstandardized variables.

3.2 Transforming Variables into Standardized Variables

Any variable Y can be transformed into a standardized variable by means of the following linear transformation:

$$z = \frac{Y - \bar{Y}}{s_Y} \tag{3.1}$$

It is easy to see that the mean of z is equal to zero because we know that $\Sigma(Y - \bar{Y})$ is equal to zero, and when we sum (3.1) we obtain

$$\Sigma z = \frac{1}{s_Y} \Sigma(Y - \bar{Y}) = 0$$

Then, with $\bar{z}$ equal to zero, we have as the variance of z

$$s_z^2 = \frac{\Sigma(z - \bar{z})^2}{n - 1} = \frac{\Sigma z^2}{n - 1}$$

or

$$s_z^2 = \frac{1}{s_Y^2} \frac{\Sigma(Y - \bar{Y})^2}{n - 1}$$

But as we have shown earlier, $s_Y^2 = \Sigma(Y - \bar{Y})^2/(n - 1)$, and, consequently,

$$s_z^2 = \frac{1}{s_Y^2} s_Y^2 = 1.0$$

3.3 The Regression Equation with Standardized Variables

Let us assume that we have n paired values of (X, Y) and that both X and Y have been transformed into standardized variables by means of (3.1). Recall now that the covariance of X and Y can be written as

$$c_{XY} = r s_X s_Y$$

and with standardized variables

$$c_{z_X z_Y} = r s_{z_X} s_{z_Y}$$

But if X and Y have been transformed into standardized variables, then $s_{z_X} = s_{z_Y} = 1$, and

$$c_{z_X z_Y} = \frac{\Sigma z_X z_Y}{n - 1} = r \qquad (3.2)$$

Similarly, the regression coefficient, $b = c_{XY}/s_X^2$, with standardized variable becomes

$$b = \frac{c_{z_X z_Y}}{s_{z_X}^2} = r$$

because $s_{z_X}^2 = 1$. Then with standardized variables the regression equation

$$Y' = \bar{Y} + b(X - \bar{X})$$

becomes

$$z'_Y = \bar{z}_Y + r(z_X - \bar{z}_X)$$

or

$$z'_Y = rz_X \tag{3.3}$$

because both $\bar{z}_Y$ and $\bar{z}_X$ are equal to zero.

3.4 Partitioning the Total Sum of Squares

In the preceding chapter we showed that

$$\Sigma(Y - \bar{Y})^2 = \Sigma(Y' - \bar{Y})^2 + \Sigma(Y - Y')^2$$
$$= \frac{(\Sigma xy)^2}{\Sigma x^2} + \left[\Sigma y^2 - \frac{(\Sigma xy)^2}{\Sigma x^2}\right]$$

Substituting in this expression with standardized variables, we have

$$\Sigma z_Y^2 = \frac{(\Sigma z_X z_Y)^2}{\Sigma z_X^2} + \left[\Sigma z_Y^2 - \frac{(\Sigma z_X z_Y)^2}{\Sigma z_X^2}\right] \tag{3.4}$$

We know that

$$s_{z_X}^2 = \frac{\Sigma z_X^2}{n-1} = s_{z_Y}^2 = \frac{\Sigma z_Y^2}{n-1} = 1$$

and, consequently,

$$\Sigma z_X^2 = \Sigma z_Y^2 = n - 1$$

Then, substituting with $n - 1$ in (3.4), we have

$$n - 1 = \frac{(\Sigma z_X z_Y)^2}{n-1} + \left[(n-1) - \frac{(\Sigma z_X z_Y)^2}{n-1}\right]$$

and dividing both sides by $n - 1$, we obtain

$$1 = r^2 + (1 - r^2) \tag{3.5}$$

We see that even though we have transformed X and Y into standardized variables, the total sum of squares can still be partitioned into a sum of squares for regression and a residual sum of squares; that r^2 is simply the proportion of the total sum of squares that can be accounted for by linear regression; and that $1 - r^2$ is the proportion that is independent of linear regression.

The transformation defined by (3.1) is a linear transformation. The correlation coefficient will remain invariant, that is, unchanged in value, by any linear transformation of either the X variable or the Y variable or of both variables.

3.5 The Partial Correlation Coefficient

Suppose that for a sample of n individuals we have obtained the following measurements: X_1 = first-year college grades, X_2 = high-school grades, and X_3 = scores on all aptitude test. We find that the correlation between first-year college grades and high-school grades is $r_{12} = .60$. We also find that the correlation between college grades and scores on the aptitude test is $r_{13} = .50$. For high-school grades and scores on the aptitude test we have $r_{23} = .50$. Now X_1 has $r_{13}^2 = (.50)^2 = .25$ of its variance in common with X_3, and X_2 has $r_{23}^2 = (.50)^2 = .25$ of its variance in common with X_3. To what degree is the correlation between X_1 and X_2 influenced by the fact that both variables have variance in common with X_3? One way to find out would be to obtain a sample in which X_3 remains constant while X_1 and X_2 vary. If all of the individuals in the sample have exactly the same aptitude test score, then for this sample, r_{13} and r_{23} would be equal to zero. Obviously, it would be a rather difficult task to obtain a sample of subjects such that they all have exactly the same aptitude test score.

A second way to find out is to obtain the *partial correlation coefficient*

$$r_{12.3} = \frac{r_{12} - r_{13}r_{23}}{\sqrt{1 - r_{13}^2}\sqrt{1 - r_{23}^2}} \tag{3.6}$$

where $r_{12.3}$ indicates the correlation coefficient between X_1 and X_2 with X_3 held constant.

In our example we have

$$r_{12.3} = \frac{.60 - (.50)(.50)}{\sqrt{1 - (.50)^2}\sqrt{1 - (.50)^2}} = .47$$

The partial correlation coefficient defined by (3.6) is the correlation coefficient between two sets of residuals that, in turn, can be shown to be uncorrelated with X_3. For the proof of (3.6), without any loss in generality, we put all three variables into standardized form. Then in standardized form the regression equation for predicting X_1 from X_3 will be

$$z_1' = r_{13}z_3$$

and the residual will be

$$z_1 - z_1' = z_1 - r_{13}z_3$$

These residuals will be uncorrelated with X_3 or z_3 because the product sum

$$\Sigma(z_1 - r_{13}z_3)z_3 = \Sigma z_1 z_3 - r_{13}\Sigma z_3^2 = r_{13}(n - 1) - r_{13}(n - 1) = 0$$

Similarly, the regression equation for predicting X_2 from X_3 in standardized form will be

$$z'_2 = r_{23}z_3$$

and the residuals $z_2 - z'_2 = z_2 - r_{23}z_3$ will be uncorrelated with z_3 or X_3. Thus, we have two sets of residuals, one for z_1 and one for z_2, which are uncorrelated with z_3 and consequently have no variance in common with X_3. The covariance between these two residuals will be

$$\frac{\Sigma(z_1 - r_{13}z_3)(z_2 - r_{23}z_3)}{n - 1} = r_{12} - r_{13}r_{23}$$

We also have as the variance of $z_1 - r_{13}z_3$

$$\frac{\Sigma(z_1 - r_{13}z_3)^2}{n - 1} = 1 - r_{13}^2$$

and as the variance of $z_2 - r_{23}z_3$

$$\frac{\Sigma(z_2 - r_{23}z_3)^2}{n - 1} = 1 - r_{23}^2$$

Then the correlation coefficient between the two sets of residuals will be

$$r_{(z_1 - r_{13}z_3)(z_2 - r_{23}z_3)} = \frac{r_{12} - r_{13}r_{23}}{\sqrt{1 - r_{13}^2}\sqrt{1 - r_{23}^2}} = r_{12.3}$$

3.6 The Semipartial Correlation Coefficient

The partial correlation coefficient removes the variance of X_3 from both X_1 and X_2. It answers the question, What is the correlation coefficient between two variables X_i and X_j when the influence of a third variable X_k is held constant? A *semipartial correlation coefficient* is the correlation coefficient between two variables X_i and X_j after the variance that X_k has in common with X_i and X_j is removed from only one of the two variables. The semipartial correlation coefficient between X_1 and X_2 with the variance of X_3 removed from X_2 will be given by

$$r_{1(2.3)} = \frac{r_{12} - r_{13}r_{23}}{\sqrt{1 - r_{23}^2}} \tag{3.7}$$

where the notation $r_{1(2.3)}$ indicates that it is the correlation coefficient between X_1 and X_2 after the variance that X_3 has in common with X_2 has been eliminated from X_2.*

* In some books the correlation coefficient defined by (3.7) is called a *part correlation* rather than a semipartial correlation.

With $r_{12} = .60$, $r_{13} = .50$, and $r_{23} = .50$ we have

$$r_{1(2.3)} = \frac{.60 - (.50)(.50)}{\sqrt{1 - (.50)^2}} = .40$$

and this semipartial correlation coefficient is the correlation between college grades (X_1) and high-school grades (X_2) after the variance of aptitude test scores (X_3) has been removed from high-school grades. Note that the variance of aptitude test scores has not been removed from college grades.

The proof of (3.7) is similar to the proof of the partial correlation coefficient. We obtain the residuals

$$z_2 - z_2' = z_2 - r_{23}z_3$$

These residuals will be uncorrelated with X_3. We do not want to eliminate the variance in common between X_1 and X_3, as we would in the case of a partial correlation coefficient. Instead, we want the semipartial correlation coefficient

$$r_{z_1(z_2 - r_{23}z_3)} = r_{1(2.3)}$$

For the covariance between z_1 and $z_2 - r_{23}z_3$ we have

$$\frac{\Sigma z_1(z_2 - r_{23}z_3)}{n - 1} = r_{12} - r_{13}r_{23}$$

The standard deviations of z_1 and $z_2 - r_{23}z_3$ will be

$$\sqrt{\frac{\Sigma z_1^2}{n - 1}} = 1 \quad \text{and} \quad \sqrt{\frac{\Sigma(z_2 - r_{23}z_3)^2}{n - 1}} = \sqrt{1 - r_{23}^2}$$

respectively. Then the correlation coefficient between z_1 and $z_2 - r_{23}z_3$ will be

$$r_{z_1(z_2 - r_{23}z_3)} = \frac{r_{12} - r_{13}r_{23}}{\sqrt{1 - r_{23}^2}} = r_{1(2.3)}$$

It is of some importance to note that the numerators of the partial correlation coefficient and the semipartial correlation coefficient are the same. Consequently, if either coefficient is equal to zero, then so must be the other.

Exercises

3.1 Can the variance of z_1' ever be larger than 1.00? Explain why or why not.

3.2 If X_1 and X_2 are transformed into standardized variables, we have $z_1' = rz_2$. Will the correlation coefficient between z_1' and X_2 be equal to the correlation coefficient between X_1 and X_2? Explain why or why not.

3.3 Will the correlation coefficient between z'_1 and X_1 be equal to the correlation coefficient between X_1 and X_2? Explain why or why not.

3.4 Can the semipartial correlation coefficient $r_{1(2.3)}$ ever be larger than the partial correlation coefficient $r_{12.3}$ for the same three variables? Explain why or why not.

3.5 Is it ever possible for the semipartial correlation $r_{1(2.3)}$ to be equal to the partial correlation coefficient $r_{12.3}$? Explain why or why not.

3.6 Why does the regression equation $Y' = a + bX$ become $z'_Y = rz_X$ if both X and Y are standardized variables?

3.7 Explain in words the meaning of the partial correlation coefficient $r_{12.3}$ and the semipartial correlation coefficient $r_{1(2.3)}$. How do these two coefficients differ?

3.8 In general, would you expect r_{12} to be larger than $r_{12.3}$? Explain why or why not.

3.9 Prove that $z_2 - z'_2$ is uncorrelated with z_1 if z'_2 is equal to $r_{12}z_1$.

3.10 If we know that the correlation coefficient between X_1 and X_2 is equal to .70, what additional information, if any, will we need in order to calculate the regression coefficient b?

3.11 If r is equal to .80, what is the value of the variance of $z_1 - z'_1$?

3.12 Will the variance of $z_1 - z'_1$ always be equal to the variance of $z_2 - z'_2$ for the same set of paired (z_1, z_2) values? Explain why or why not.

3.13 Will the variance of z'_1 be equal to the variance of z'_2 for the same set of paired (z_1, z_2) values? Explain why or why not.

3.14 Is it ever possible for $r_{12.3}$ to be larger than r_{12}? Explain why or why not.

3.15 If $r_{12} = .70$ and $r_{13} = .60$, do these two values set a limit on the possible values of r_{23}? If so, what are these limits?

3.16 Using the limits found in Exercise 3.15, what are the values of the semipartial correlation coefficient $r_{1(2.3)}$?

3.17 Explain each of the following concepts:

partial correlation coefficient

standardized variables

semipartial correlation coefficient

4

Multiple Regression and Correlation

4.1 Introduction

In this chapter we examine the correlation coefficient between a variable of primary interest, Y, and a weighted linear sum of a number of X variables. This correlation coefficient is called a *multiple correlation coefficient* and is customarily indicated by R. The basic principles are the same as when we had a single X variable, but instead of a regression equation involving a single X variable, we will have a multiple regression equation

$$Y' = a + b_1X_1 + b_2X_2 + \cdots + b_kX_k$$

and we want to find the values of $a, b_1, b_2, \ldots, b_k$ that will result in the highest possible positive correlation coefficient between the observed values of Y and the predicted values Y'. When this is done, the resulting correlation coefficient is called a multiple correlation coefficient and is represented by either $R_{YY'}$ or $R_{Y.12\ldots k}$, or simply by R.

As the number of X variables increases, the calculations involved in finding the values of $b_1, b_2, \ldots, b_k$ become complex and overwhelming with a hand calculator, although they can be accomplished quite easily by a high-speed electronic computer. The principles of multiple regression and correlation can be illustrated, however, with an example consisting of one Y variable and two X variables. With only two X variables the calculations are relatively simple.

4.2 The Least Squares Equations for *a* and the Two Regression Coefficients

With two X variables, the multiple regression equation is

$$Y' = a + b_1X_1 + b_2X_2 \tag{4.1}$$

As in the case of a single X variable, the total sum of squares for the Y variable can be partitioned into two orthogonal components, or

$$\begin{aligned} SS_{tot} &= SS_{reg} + SS_{res} \\ &= \Sigma(Y' - \bar{Y})^2 + \Sigma(Y - Y')^2 \end{aligned}$$

and we want to find the values of a, b_1, and b_2 that will minimize the residual sum of squares, $\Sigma(Y - Y')^2$. If we minimize $\Sigma(Y - Y')^2$, we will at the same time maximize the regression sum of squares, $\Sigma(Y' - \bar{Y})^2$.

It can be shown that the values of a, b_1, and b_2 that will minimize $\Sigma(Y - Y')^2$ must satisfy the following equations:*

$$an + b_1\Sigma X_1 + b_2\Sigma X_2 = \Sigma Y \tag{4.2a}$$

$$a\Sigma X_1 + b_1\Sigma X_1^2 + b_2\Sigma X_1X_2 = \Sigma X_1Y \tag{4.2b}$$

$$a\Sigma X_2 + b_1\Sigma X_2X_1 + b_2\Sigma X_2^2 = \Sigma X_2Y \tag{4.2c}$$

Although we can solve the above three equations with three unknowns, a, b_1, and b_2, by standard algebraic methods, it is considerably

---

* The proof of these equations can be obtained by expanding the right side of

$$\Sigma(Y - Y')^2 = \Sigma(Y - a - b_1X_1 - b_2X_2)^2$$

and differentiating with respect to a, b_1, and b_2. Setting these derivatives equal to zero, we obtain equations (4.2a), (4.2b), and (4.2c).

easier to solve two equations with two unknowns by these methods. Let us see, therefore, if we can eliminate one of the three equations. Suppose, for example, that we express X_1, X_2, and Y in deviation form. Then, because $\Sigma x_1 = \Sigma x_2 = \Sigma y = 0$, it is obvious that (4.2a) will be equal to zero or

$$an + b_1\Sigma x_1 + b_2\Sigma x_2 = \Sigma y = 0 \tag{4.3}$$

This leaves us with (4.2b) and (4.2c) and with $\Sigma x_1 = \Sigma x_2 = 0$; these two equations become

$$b_1\Sigma x_1^2 + b_2\Sigma x_1 x_2 = \Sigma x_1 y \tag{4.4}$$

and

$$b_1\Sigma x_2 x_1 + b_2\Sigma x_2^2 = \Sigma x_2 y \tag{4.5}$$

The above two equations with two unknowns, b_1 and b_2, can be solved fairly easily. For example, if we multiply (4.4) by Σx_2^2 and (4.5) by $\Sigma x_1 x_2$, we obtain

$$b_1(\Sigma x_1^2)(\Sigma x_2^2) + b_2(\Sigma x_1 x_2)(\Sigma x_2^2) = (\Sigma x_1 y)(\Sigma x_2^2) \tag{4.6}$$

and

$$b_1(\Sigma x_1 x_2)^2 + b_2(\Sigma x_1 x_2)(\Sigma x_2^2) = (\Sigma x_2 y)(\Sigma x_1 x_2) \tag{4.7}$$

Then, subtracting (4.7) from (4.6), we have

$$b_1(\Sigma x_1^2)(\Sigma x_2^2) - b_1(\Sigma x_1 x_2)^2 = (\Sigma x_1 y)(\Sigma x_2^2) - (\Sigma x_2 y)(\Sigma x_1 x_2)$$

and

$$b_1 = \frac{(\Sigma x_1 y)(\Sigma x_2^2) - (\Sigma x_2 y)(\Sigma x_1 x_2)}{(\Sigma x_1^2)(\Sigma x_2^2) - (\Sigma x_1 x_2)^2} \tag{4.8}$$

Following a similar procedure, we obtain

$$b_2 = \frac{(\Sigma x_2 y)(\Sigma x_1^2) - (\Sigma x_1 y)(\Sigma x_1 x_2)}{(\Sigma x_1^2)(\Sigma x_2^2) - (\Sigma x_1 x_2)^2} \tag{4.9}$$

4.3 An Example with Two *X* Variables

Table 4.1 gives the values of X_1, X_2, and Y for a sample of $n = 25$ observations. Table 4.2 summarizes the results of calculations involving these three variables. The means, $\bar{X}_1$, $\bar{X}_2$, and $\bar{Y}$, are shown at the bottom of the table, and the standard deviations appear at the right of the table. The diagonal entries in the table are the sums of squared deviations from the means. For example, we have $\Sigma x_1^2 = 217.36$. The nondiagonal entries give the sums of the products of the paired deviations. For example, we have $\Sigma x_1 x_2 = 232.56$. Substituting the appro-

TABLE 4.1 Values of X_1, X_2, and Y for a sample of 25 subjects

Subjects	X_1	X_2	Y
1	11	38	10
2	7	42	16
3	12	38	18
4	13	36	15
5	14	40	15
6	15	32	11
7	5	20	13
8	14	44	18
9	14	34	12
10	10	28	16
11	8	24	10
12	16	30	16
13	15	26	15
14	14	24	12
15	10	26	12
16	9	18	14
17	11	30	16
18	9	26	13
19	7	18	11
20	10	10	17
21	9	12	8
22	10	32	18
23	10	18	14
24	16	20	15
25	10	18	12

TABLE 4.2 Means, standard deviations, sums of squared deviations and sums of products of deviations for the observations in Table 4.1

Variable	X_1	X_2	Y	Standard deviations
X_1	217.36	232.56	47.48	3.0094
X_2		2077.77	230.08	9.3045
Y			180.64	2.7435
Means	11.16	27.36	13.88	

priate values from Table 4.2 in (4.8) and (4.9), we obtain

$$b_1 = \frac{(47.48)(2077.77) - (230.08)(232.56)}{(217.36)(2077.77) - (232.56)^2} = .1136$$

and

$$b_2 = \frac{(230.08)(217.36) - (47.48)(232.56)}{(217.36)(2077.77) - (232.56)^2} = .0980$$

Dividing (4.2a) by n and solving for a, we have

$$a = \bar{Y} - b_1\bar{X}_1 - b_2\bar{X}_2$$

and

$$a = 13.88 - (.1136)(11.16) - (.0980)(27.36) = 9.9309$$

Then the multiple regression equation will be

$$Y' = 9.9309 + .1136X_1 + .0980X_2$$

The meaning of a regression coefficient in a multiple regression equation is simply this: If all variables except X_i are held constant, then b_i is the amount by which Y' increases with unit increase in X_i. In our example, $b_1 = .1136$ is the amount by which Y' will increase with unit increase in X_1 if X_2 is held constant, and $b_2 = .0980$ is the amount by which Y' will increase with unit increase in X_2 if X_1 is held constant.*

We have previously shown, in the case of two variables, X and Y, that the total sum of squares for the Y variable, or

$$SS_{tot} = \Sigma y^2 = \Sigma(Y - \bar{Y})^2$$

can be partitioned into two orthogonal components, the sum of squares for regression, or

$$SS_{reg} = b\Sigma xy = \frac{(\Sigma xy)^2}{\Sigma x^2} = \Sigma(Y' - \bar{Y})^2$$

and the residual sum of squares, or

$$SS_{res} = \Sigma(Y - Y')^2$$

This can also be done in a multiple regression problem involving two or more X variables; that is, we will also have

$$SS_{tot} = SS_{reg} + SS_{res}$$

or

$$\Sigma(Y - \bar{Y})^2 = \Sigma(Y' - \bar{Y})^2 + \Sigma(Y - Y')^2$$

* In a multiple regression analysis with more than one X variable, the regression coefficients are commonly referred to as partial regression coefficients and are designated by $b_{1.2}$ and $b_{2.1}$ in the case of two X variables. With three X variables the regression coefficients would be designated by $b_{1.23}$, $b_{2.13}$, and $b_{3.12}$. The dot notation is necessary only to indicate that $b_{1.23}$ is not the same as b_1 in a regression with X_1 alone or $b_{1.2}$ in a regression with X_1 and X_2. Because we shall be concerned only with partial regression coefficients when all X variables are in the regression equation, we use the simpler notation b_1 for $b_{1.23\ldots k}$.

In a multiple regression problem, the regression sum of squares will be given by

$$SS_{reg} = b_1\Sigma x_1 y + b_2\Sigma x_2 y + \cdots + b_k\Sigma x_k y \tag{4.10}$$

In our example we have two values of b with $b_1 = .1136$ and $b_2 = .0980$. From Table 4.2 we see that $\Sigma x_1 y = 47.48$ and $\Sigma x_2 y = 230.08$. Then we have

$$SS_{reg} = (.1136)(47.48) + (.0980)(230.08) = 27.94$$

as the sum of squares for linear regression with $k = 2$ d.f.* Then the residual sum of squares can be obtained by subtraction, that is,

$$SS_{res} = SS_{tot} - SS_{reg}$$

and in our example we have

$$SS_{res} = 180.64 - 27.94 = 152.70$$

with $n - k - 1 = 22$ d.f.

The square of the multiple correlation coefficient will then be given by

$$R^2_{YY'} = \frac{SS_{reg}}{SS_{tot}} \tag{4.11}$$

and in our example we have

$$R^2_{YY'} = \frac{27.94}{180.64} = .1547$$

The square root of $R^2_{YY'}$ is, of course, the multiple correlation coefficient, and

$$R_{YY'} = \sqrt{.1547} = .3933$$

We also note that

$$SS_{tot}R^2_{YY'} = SS_{reg} \tag{4.12}$$

and that

$$1 - R^2_{YY'} = 1 - \frac{SS_{reg}}{SS_{tot}} = \frac{SS_{tot} - SS_{reg}}{SS_{tot}} = \frac{SS_{res}}{SS_{tot}}$$

Then we also have

$$SS_{tot}(1 - R^2_{YY'}) = SS_{res} \tag{4.13}$$

---

* Recall from an earlier discussion that the degrees of freedom for SS_{reg} will be equal to k, where k is the number of X variables, and the degrees of freedom for SS_{res} will be equal to $n - k - 1$, where n is the total number of observations.

and

$$\frac{SS_{tot}}{SS_{tot}} = \frac{SS_{reg}}{SS_{tot}} + \frac{SS_{res}}{SS_{tot}}$$

or

$$1 = R^2_{YY'} + (1 - R^2_{YY'})$$

Thus we see that $R^2_{YY'}$ is the proportion of SS_{tot} that can be accounted for by the linear regression of Y on Y' and $1 - R^2_{YY'}$ is the proportion of SS_{tot} that is independent of or that cannot be accounted for by the regression of Y on Y'.

4.4 Calculating $R^2_{Y.12}$ and b_1 and b_2 Using Correlation Coefficients

The correlation coefficients between Y and X_1, Y and X_2, and X_1 and X_2 can be obtained from the data given in Table 4.2. For example, we have

$$r_{Y1} = \frac{\Sigma x_1 y}{\sqrt{\Sigma x_1^2}\sqrt{\Sigma y^2}} = \frac{47.48}{\sqrt{217.36}\sqrt{180.64}} = .2396$$

$$r_{Y2} = \frac{\Sigma x_2 y}{\sqrt{\Sigma x_2^2}\sqrt{\Sigma y^2}} = \frac{230.08}{\sqrt{2077.77}\sqrt{180.64}} = .3756$$

and

$$r_{12} = \frac{\Sigma x_1 x_2}{\sqrt{\Sigma x_1^2}\sqrt{\Sigma x_2^2}} = \frac{232.56}{\sqrt{217.36}\sqrt{2077.77}} = .3461$$

Then $R^2_{YY'}$ with two X variables will also be given by

$$R^2_{Y.12} = \frac{r^2_{Y1} + r^2_{Y2} - 2r_{Y1}r_{Y2}r_{12}}{1 - r^2_{12}} \tag{4.14}$$

and in our example we have

$$R^2_{Y.12} = \frac{(.2396)^2 + (.3756)^2 - 2(.2396)(.3756)(.3461)}{1 - (.3461)^2} = .1547$$

which is equal to the value we obtained before.

The regression coefficients, b_1 and b_2, can also be obtained using the correlation coefficients. For example, if we divide both the numerator and the denominator of (4.8) by $(n - 1)^2$ and simplify the resulting expression, we obtain

$$b_1 = \frac{s_Y(r_{Y1} - r_{Y2}r_{12})}{s_{X_1}(1 - r^2_{12})} \tag{4.15}$$

Similarly, dividing both the numerator and the denominator of (4.9) by $(n-1)^2$, we obtain

$$b_2 = \frac{s_Y(r_{Y2} - r_{Y1}r_{12})}{s_{X_2}(1 - r_{12}^2)} \tag{4.16}$$

Substituting the appropriate values in (4.15) and (4.16), we have

$$b_1 = \frac{2.7435[.2396 - (.3756)(.3461)]}{3.0094[1 - (.3461)^2]} = .1135$$

and

$$b_2 = \frac{2.7435[.3756 - (.2396)(.3461)]}{9.3045[1 - (.3461)^2]} = .0980$$

which are equal, within rounding errors, to the values we obtained before.

If X_1, X_2, and Y are in standardized form, then the variance and standard deviation of these standardized variables will be equal to 1. We note that the regression coefficients defined by (4.15) and (4.16) will therefore be equal to

$$\hat{b}_1 = \frac{r_{Y1} - r_{Y2}r_{12}}{1 - r_{12}^2} \tag{4.17}$$

and

$$\hat{b}_2 = \frac{r_{Y2} - r_{Y1}r_{12}}{1 - r_{12}^2} \tag{4.18}$$

respectively. We use $\hat{b}$ to indicate the regression coefficients for variables in standardized form. We can, of course, obtain b_1 and b_2 from the values of $\hat{b}_1$ and $\hat{b}_2$, because

$$b_1 = \frac{s_Y}{s_{X_1}}\hat{b}_1 \quad \text{and} \quad b_2 = \frac{s_Y}{s_X{}^2}\hat{b}_2$$

4.5 The Semipartial Correlation Coefficients $r_{Y(1.2)}$ and $r_{Y(2.1)}$

With both X_1 and X_2 in the regression equation, the sum of squares for linear regression is $SS_{reg} = 27.94$ with 2 d.f. Now suppose we find the sum of squares for linear regression when only X_1 is in the regression equation. In this instance we have

$$SS_{reg_1} = \frac{(\Sigma x_1 y)^2}{\Sigma x_1^2} = \frac{(47.48)^2}{217.36} = 10.37$$

with 1 d.f. Note also that this sum of squares will be equal to

$$SS_{reg_1} = r_{Y1}^2 SS_{tot} = (.2396)^2(180.64) = 10.37$$

The difference, $27.94 - 10.37 = 17.57$, will represent the increment in the regression sum of squares when X_2 is included in the regression equation. This increment when divided by the total sum of squares is the square of the semipartial correlation coefficient, $r^2_{Y(2.1)}$, and

$$r^2_{Y(2.1)} = \frac{17.57}{180.64} = .0973$$

and $r_{Y(2.1)} = \sqrt{.0973} = .3119$.

In Section 3.6 we showed that the semipartial correlation $r_{Y(2.1)}$ is given by

$$r_{Y(2.1)} = \frac{r_{Y2} - r_{Y1}r_{12}}{\sqrt{1 - r^2_{12}}}$$

and substituting the appropriate values in the formula for $r_{Y(2.1)}$, we have

$$r_{Y(2.1)} = \frac{.3756 - (.2396)(.3461)}{\sqrt{1 - (1.3461)^2}} = .3119$$

and $r^2_{Y(2.1)} = (.3119)^2 = .0973$.

The semipartial correlation $r_{Y(2.1)}$, as we pointed out in Section 3.6, is simply the correlation between X_2 and Y after the variance that X_1 has in common with X_2 has been removed from X_2. We know that $r^2_{Y1} = (.2396)^2 = .0574$ is the proportion of $\Sigma(Y - \bar{Y})^2$ that can be accounted for by the regression of Y on X_1. Given that X_1 has been included in the regression equation, the additional proportion of $\Sigma(Y - \bar{Y})^2$ that can be accounted for by X_2 will be given by $r^2_{Y(2.1)} = (.3119)^2 = .0973$. We now note that

$$R^2_{Y.12} = r^2_{Y1} + r^2_{Y(2.1)}$$

or, in our example,

$$R^2_{Y.12} = .0574 + .0973 = .1547$$

In the same manner we could show that the semipartial correlation $r_{Y(1.2)}$ will be given by

$$r_{Y(1.2)} = \frac{r_{Y1} - r_{Y2}r_{12}}{\sqrt{1 - r^2_{12}}}$$

or, in our example,

$$r_{Y(1.2)} = \frac{.2396 - (.3756)(.3461)}{\sqrt{1 - (.3461)^2}} = .1168$$

and $r^2_{Y(1.2)} = (.1168)^2 = .0136$.

The regression sum of squares accounted for by X_2 alone is

$$SS_{reg_2} = \frac{(\Sigma x_2 y)^2}{\Sigma x_2^2} = \frac{(230.08)^2}{2077.70} = 25.48$$

which is also equal to

$$SS_{reg_2} = r_{Y2}^2 SS_{tot} = (.3756)^2(180.64) = 25.48$$

Given that X_2 is in the regression equation, the increment in the regression sum of squares that can be accounted for by X_1 is 27.94 − 25.48 = 2.46, and we also have

$$r_{Y(1.2)}^2 = \frac{2.46}{180.64} = .0136$$

which is equal to the value we obtained before. Again we note that

$$R_{Y.12}^2 = r_{Y2}^2 + r_{Y(1.2)}^2$$

or, in our example,

$$R_{Y.12}^2 = (.3756)^2 + (.1168)^2 = .1547$$

It is also important to observe that the squared semipartial correlations, $r_{Y(1.2)}^2$ and $r_{Y(2.1)}^2$, can also be obtained by means of the following equations:

$$r_{Y(1.2)}^2 = R_{Y.12}^2 - r_{Y2}^2$$

and

$$r_{Y(2.1)}^2 = R_{Y.12}^2 - r_{Y1}^2$$

With $k > 2X$ variables, these equations generalize to the difference between two squared multiple correlation coefficients.* For example, with three X variables we would have

$$r_{Y(1.23)}^2 = R_{Y.123}^2 - R_{Y.23}^2$$
$$r_{Y(2.13)}^2 = R_{Y.123}^2 - R_{Y.13}^2$$
$$r_{Y(3.12)}^2 = R_{Y.123}^2 - R_{Y.12}^2$$

Then consider any one, say the last, of the above three equations. We have shown that

$$R_{Y.12}^2 = r_{Y1}^2 + r_{Y(2.1)}^2$$

so that

$$r_{Y(3.12)}^2 = R_{Y.123}^2 - r_{Y1}^2 - r_{Y(2.1)}^2$$

* Squared partial correlation coefficients can also be obtained from the values of squared multiple correlation coefficients. For example,

$$r_{Y1.23}^2 = \frac{R_{Y.123}^2 - R_{Y.23}^2}{1 - R_{Y.23}^2}$$

ıerefore,

$$R^2_{Y.123} = r^2_{Y1} + r^2_{Y(2.1)} + r^2_{Y(3.12)} \tag{4.19}$$

Equation (4.19) generalizes to any number of X variables; that is,

$$R^2_{Y.123\ldots k} = r^2_{Y1} + r^2_{Y(2.1)} + r^2_{Y(3.12)} + \cdots + r^2_{Y(k.123\ldots k-1)}$$

For three X variables the proportion of the total sum of squares accounted for by X_1, given that it is entered first in the regression equation, will be r^2_{Y1}. After X_1 is entered in the regression equation, the increment in the proportion of the total sum of squares accounted for by X_2, given that it is entered second in the regression equation, will be $r^2_{Y(2.1)}$. Now that both X_1 and X_2 are in the regression equation, the increment due to X_3 will be given by $r^2_{Y(3.12)}$.

We emphasize that the proportion of $\Sigma(Y - \bar{Y})^2$ accounted for by a given variable depends on the position in which that variable is entered into the regression equation. In our example, when X_1 is entered first it accounts for $r^2_{Y1} = (.2396)^2 = .0574$ of the total sum of squares. But given that X_2 is entered first, then X_1 accounts for $r^2_{Y(1.2)} = (.1168)^2 = .0136$ of the total sum of squares. Similarly, if X_2 is entered first in the regression equation, it accounts for $r^2_{Y2} = (.3756)^2 = .1411$ of the total sum of squares. If X_2 is entered after X_1, it accounts for $r^2_{Y(2.1)} = (.3119)^2 = .0973$ of the total sum of squares.

When X variables are intercorrelated, there is really no satisfactory method of determining the relative contribution of the X variables to the regression sum of squares or the proportion of $\Sigma(Y - \bar{Y})^2$ accounted for by each of the X variables. This will depend, as we have just seen, on the order in which the X variables are entered in the regression equation. When X variables are correlated, each variable usually accounts for a larger proportion of $\Sigma(Y - \bar{Y})^2$ when it is entered first in the regression equation than when it follows other variables.

4.6 Multiple Regression with Orthogonal X variables

w/ true orthogonal variables

If the intercorrelation of the X variables in a multiple regression analysis are all equal to zero, then any semipartial correlation, say $r_{Y(3.12)}$, will simply be equal to r_{Y3}. This will be true for each and every semipartial correlation that we might calculate. In this instance we would have

$$R^2_{Y.123\ldots k} = r^2_{Y1} + r^2_{Y2} + r^2_{Y3} + \cdots + r^2_{Yk}$$

Furthermore, the regression coefficients $b_1, b_2, b_3, \ldots, b_k$ would remain exactly the same even though we might drop several of the X variables from the regression equation. In addition, the proportion of SS_{tot}

accounted for by each X variable would remain the same, no matter in what sequence the X variables were entered into the regression equation. And the proportion of SS_{tot} accounted for by each X variable would remain the same even though some of the X variables might be dropped from the regression equation.

4.7 Test of Significance of $R_{YY'}$

As (4.10) shows, if both b_1 and b_2 are equal to zero, then SS_{reg} will be equal to zero and $R_{YY'}$ will also be equal to zero. The obtained values of b_1 and b_2 are estimates of the corresponding population values, β_1 and β_2. A test of the null hypothesis that the two population values, β_1 and β_2, are both equal to zero will be given by

$$F = \frac{R^2_{YY'}/k}{(1 - R^2_{YY'})/(n - k - 1)} \tag{4.20}$$

where k is the number of X variables. If both the numerator and the denominator of (4.20) are multiplied by SS_{tot}, we have the equivalent test,

$$F = \frac{SS_{reg}/k}{SS_{res}/(n - k - 1)} = \frac{MS_{reg}}{MS_{res}} \tag{4.21}$$

For both F ratios the degrees of freedom for the numerator will be equal to k, the number of X variables, and the degrees of freedom for the denominator will be equal to $n - k - 1$.

With (4.20) we have, in our example,

$$F = \frac{.1547/2}{(1 - .1547)/22} = 2.01$$

or, equivalently, with (4.21),

$$F = \frac{27.94/2}{152.70/22} = 2.01$$

with 2 and 22 d.f. With $\alpha = .05$, this is not a significant value of F.

4.8 Test of Significance of b_1 and b_2 and the Equivalent Test of Significance of $r_{Y(1.2)}$ and $r_{Y(2.1)}$

For a test of significance of a regression coefficient in a multiple regression equation, we have the t test, or

$$t = \frac{b_i}{\sqrt{\dfrac{MS_{res}}{\Sigma x_i^2(1 - R^2_{i.})}}} \tag{4.22}$$

where $R_{i.}^2$ is the squared multiple correlation of X_i with the remaining X variables. With only two X variables, $R_{i.}^2 = r_{12}^2$. The square of t will be an F ratio with 1 and $n - k - 1$ degrees of freedom. As an identity for b_1, given by (4.15), we have

$$b_1 = \frac{s_Y(r_{Y1} - r_{Y2}r_{12})}{s_{X_1}(1 - r_{12}^2)}$$

It will be much more insightful to examine the F test for b_1 rather than the t test, and we shall do so. Then, substituting the value of b_1 as given above in (4.22) and squaring, we have

$$F = \frac{\dfrac{\Sigma y^2(r_{Y1} - r_{Y2}r_{12})^2/(n-1)}{\Sigma x_1^2(1 - r_{12}^2)^2/(n-1)}}{MS_{res}/(\Sigma x_1^2)(1 - r_{12}^2)} \tag{4.23}$$

and canceling terms, we obtain

$$F = \frac{\Sigma y^2(r_{Y1} - r_{Y2}r_{12})^2/(1 - r_{12}^2)}{MS_{res}}$$

We have shown that

$$MS_{res} = (1 - R_{YY'}^2)\Sigma y^2/(n - k - 1)$$

and substituting this expression in (4.23) and canceling terms, we obtain

$$F = \frac{(r_{Y1} - r_{Y2}r_{12})^2/(1 - r_{12}^2)}{(1 - R_{YY'}^2)/(n - k - 1)} \tag{4.24}$$

We now observe that the numerator of (4.24) is simply the square of the semipartial correlation coefficient $r_{Y(1.2)} = (r_{Y1} - r_{Y2}r_{12})/\sqrt{1 - r_{12}^2}$, and the F test of b_1 is equivalent to the F test of the semipartial correlation coefficient $r_{Y(1.2)}$. The equivalence of these two tests applies in the case of any number of X variables. If we have, for example, four X variables, the F test of b_1 will be equivalent to the F test of the semipartial correlation coefficient $r_{Y(1.234)}$. Similarly, the F test of b_4 will be equivalent to the F test of $r_{Y(4.123)}$.

Recall that

$$r_{Y(1.234)}^2 = R_{Y.1234}^2 - R_{Y.234}^2$$

and $r_{Y(1.234)}^2$ is simply the proportion of $\Sigma(Y - \bar{Y})^2$ that can be accounted for by X_1 when it is entered last in the regression equation or that proportion of $\Sigma(Y - \bar{Y})^2$ that is not already accounted for by X_2, X_3, and X_4. Then we also have

$$F = \frac{r_{Y(1.234)}^2}{(1 - R_{Y.1234}^2)/(n - k - 1)} = \frac{R_{Y.1234}^2 - R_{Y.234}^2}{(1 - R_{Y.1234}^2)/(n - k - 1)}$$

The test of significance of $r_{Y(1.234)}$ or equivalently of b_1 is simply a test to determine whether the regression sum of squares already accounted for by X_2, X_3, and X_4 is increased significantly by including X_1 as the last variable in the regression equation. Similarly, the test of significance of $r_{Y(4.123)}$ or, equivalently, of b_4 is a test to determine whether the regression sum of squares is increased significantly by X_4 above that already accounted for by X_1, X_2, and X_3. The value of $r^2_{Y(4.123)}$ is simply the proportion of $\Sigma(Y - \bar{Y})^2$ that is accounted for by X_4 when it is entered last in the regression equation.

4.9 Plotting of Residuals

The earlier discussion regarding plots of residuals, $e = Y - Y'$, against X applies even more strongly to the situation in which we have more than one X variable. In general, if the various assumptions of the linear regression model are satisfied, a plot of the residuals against each of the X variables should show no discernible trends or patterns of the kind described in Chapter 2.

Exercises

4.1 We have $r_{Y1} = -.80$ and $r_{Y2} = .60$.
(a) What is the smallest possible value for R? Explain why.
(b) Is it possible for R to be equal to 1.00? Explain why or why not.

4.2 We have four X variables.
(a) Write a formula that will give the squared semipartial correlation coefficient $r^2_{Y(1.234)}$.
(b) Write a formula that will give the squared partial correlation coefficient $r^2_{Y1.234}$.

4.3 Explain in words the meaning of $r^2_{Y(1.234)}$ and $r^2_{Y1.234}$.

4.4 If $r_{Y1} = .70$ and $r_{Y2} = .60$, is it possible for r_{12} to be equal to $-.30$? Explain why or why not.

4.5 If $r_{Y1} = .20$, $r_{Y2} = .40$, and $r_{Y3} = .60$, and if $r_{12} = r_{13} = r_{23} = 0$, what is the value of $R_{Y.123}$?

4.6 Assume that we have $n = 20$ observations with

$\bar{X}_1 = 4.0$ $\quad \Sigma x_1^2 = 76.0$ $\quad \Sigma x_1 y = 212.8$

$\bar{X}_2 = 8.0$ $\quad \Sigma x_2^2 = 304.0$ $\quad \Sigma x_2 y = 364.8$

$\bar{Y} = 20.0$ $\quad \Sigma y^2 = 1216.0$ $\quad \Sigma x_1 x_2 = 0$

(a) Find the values of b_1 and b_2 in the multiple regression equation

$$Y' = a + b_1X_1 + b_2X_2$$

(b) Find the value of b_1 in the regression equation $Y' = a + b_1X_1$. Is this value of b_1 equal to the value of b_1 in the multiple regression equation?

(c) Is the value of b_2 in the regression equation $Y' = a + b_2X_2$ equal to the value of b_2 obtained in the multiple regression equation?

(d) Is MS_{reg} significant with $\alpha = .05$?

(e) Is $R_{Y.12}$ significant with $\alpha = .05$?

(f) What proportion of SS_{tot} is accounted for by X_1?

(g) What proportion of SS_{tot} is accounted for by X_2?

(h) What proportion of SS_{tot} is accounted for jointly by X_1 and X_2?

4.7 Assume that we have $n = 20$ observations with

$$\bar{X}_1 = 4.0 \qquad \Sigma x_1^2 = 76.0 \qquad \Sigma x_1 y = 152.0$$

$$\bar{X}_2 = 8.0 \qquad \Sigma x_2^2 = 304.0 \qquad \Sigma x_2 y = 304.0$$

$$\bar{Y} = 20.0 \qquad \Sigma y^2 = 1216.0 \qquad \Sigma x_1 x_2 = 76.0$$

(a) Find the values of r_{Y1}, r_{Y2}, and r_{12}.

(b) Find the values of b_1 and b_2 in the multiple regression equation

$$Y' = a + b_1X_1 + b_2X_2$$

(c) Is $R_{Y.12}$ significant with $\alpha = .05$?

(d) Given that X_1 has been included in the regression equation, does X_2 contribute significantly to the regression sum of squares?

(e) Given that X_2 is in the regression equation, does X_1 contribute significantly to the regression sum of squares?

(f) Is the test of significance of $R_{Y.12}$ equivalent to the test of significance of MS_{reg}; that is, would we obtain the same F ratios for both tests? Explain why or why not.

4.8 Explain why

$$r_{Y1.23}^2 = \frac{R_{Y.123}^2 - R_{Y.23}^2}{1 - R_{Y.23}^2}$$

is a squared partial correlation coefficient.

4.9 Explain the following concepts:

multiple correlation coefficient

multiple regression equation

5

Matrix Calculations for Regression Coefficients

5.1 The Matrix Product $\mathbf{R}^{-1}\mathbf{r} = \hat{\mathbf{b}}$

If k is the number of X variables in a multiple regression problem and if these variables, along with Y, are in standardized form, then the value of a in the regression equation will be equal to zero. In this instance all of the regression coefficients will be in standardized form, and the regression equation will take the form*

$$Y' = \hat{b}_1 z_1 + \hat{b}_2 z_2 + \cdots + \hat{b}_k z_k$$

If the residual sum of squares is to be minimized, then the regression coefficients must satisfy the following equations:

$$\begin{aligned}
\hat{b}_1 r_{11} + \hat{b}_2 r_{12} + \hat{b}_3 r_{13} + \cdots + \hat{b}_k r_{1k} &= r_{Y1} \\
\hat{b}_1 r_{21} + \hat{b}_2 r_{22} + \hat{b}_3 r_{23} + \cdots + \hat{b}_k r_{2k} &= r_{Y2} \\
\hat{b}_1 r_{31} + \hat{b}_2 r_{32} + \hat{b}_3 r_{33} + \cdots + \hat{b}_k r_{3k} &= r_{Y3} \\
\vdots \\
\hat{b}_1 r_{k1} + \hat{b}_2 r_{k2} + \hat{b}_3 r_{k3} + \cdots + \hat{b}_k r_{kk} &= r_{Yk}
\end{aligned} \tag{5.1}$$

The $k \times k$ intercorrelations of the X variables can be represented by a symmetric matrix that we designate by **R.** The k correlations of the X variables with the Y variable can be represented by a $k \times 1$ column vector that we designate by **r.** Similarly, the k values of the regression coefficients can be represented by a $k \times 1$ column vector that we designate by $\hat{\mathbf{b}}$.

If we can find the inverse of the **R** matrix, which we designate by $\mathbf{R}^{-1}$, then it can be shown that the matrix product is†

$$\mathbf{R}^{-1}\mathbf{r} = \hat{\mathbf{b}} \tag{5.2}$$

5.2 The Inverse of a Symmetric 2 × 2 Correlation Matrix

To differentiate between the elements of $\mathbf{R}^{-1}$ and the correlation coefficients in **R,** we represent the former by superscripts or by r^{ii} and r^{ij}. We illustrate the matrix calculations for two X variables symbolically as follows:

$$\begin{bmatrix} r^{11} & r^{12} \\ r^{21} & r^{22} \end{bmatrix} \begin{bmatrix} r_{Y1} \\ r_{Y2} \end{bmatrix} = \begin{bmatrix} \hat{b}_1 \\ \hat{b}_2 \end{bmatrix}$$

$$\mathbf{R}^{-1} \quad \mathbf{r} \quad = \quad \hat{\mathbf{b}}$$

---

* It is important to note that Y' is not necessarily a standardized variable. Although the mean of the Y' values will always be equal to zero, the variance of the Y' values will be equal to $R^2_{YY'}$. Only in the unusual case where $R^2_{YY'}$ is equal to one will Y' be equal to $z_{Y'}$. For further discussion, see Exercise 5.12.

† Although a prior knowledge of matrix algebra is not essential for understanding this and subsequent chapters, your understanding of multiple regression analysis, in general, will be considerably enhanced if you know "a little bit about matrices and matrix algebra." For a beginning, I suggest you read the chapter of the same title in Edwards (1984a). See also Horst (1963) and Searle (1982) for books that were written for social scientists rather than for mathematicians.

where

$$r^{11}r_{Y1} + r^{12}r_{Y2} = \hat{b}_1 \quad \text{and} \quad r^{21}r_{Y1} + r^{22}r_{Y2} = \hat{b}_2$$

The inverse of **R** is defined if $\mathbf{R}^{-1}\mathbf{R} = \mathbf{I}$, where **I** is an identity matrix with 1's in the upper-left to lower-right diagonal and 0's elsewhere or, in the case of a 2×2 matrix, if

$$\underset{\mathbf{R}^{-1}}{\begin{bmatrix} r^{11} & r^{12} \\ r^{21} & r^{22} \end{bmatrix}} \underset{\mathbf{R}}{\begin{bmatrix} r_{11} & r_{12} \\ r_{21} & r_{22} \end{bmatrix}} \underset{=}{=} \underset{\mathbf{I}}{\begin{bmatrix} 1 & 0 \\ 0 & 1 \end{bmatrix}}$$

where

$$r^{11}r_{11} + r^{12}r_{21} = 1$$
$$r^{11}r_{12} + r^{12}r_{22} = 0$$
$$r^{21}r_{11} + r^{22}r_{21} = 0$$
$$r^{21}r_{12} + r^{22}r_{22} = 1$$

We note that for the symmetric 2×2 correlation matrix, $r_{11} = r_{22} = 1$ and $r_{21} = r_{12}$; in this case, it can easily be shown that the elements of the inverse matrix $\mathbf{R}^{-1}$, provided that r_{12} is not equal to 1.00 or -1.00, will be given by

$$\mathbf{R}^{-1} = c\begin{bmatrix} r_{11} & -r_{12} \\ -r_{21} & r_{22} \end{bmatrix}$$

where $c = 1/(1 - r_{12}^2)$ is a scalar or constant that multiplies each of the elements in the matrix on the right.

5.3 Calculation of the Values of $\hat{b}_1$ and $\hat{b}_2$

Recall that for the example considered in the preceding chapter we had $r_{12} = r_{21} = .3461$. We also had $r_{Y1} = .2396$ and $r_{Y2} = .3756$. Then for the inverse of **R** we first obtain $c = 1/(1 - r_{12}^2) = 1/[1 - (.3461)^2] = 1/.8802 = 1.1361$, and then we calculate

$$\mathbf{R}^{-1} = 1.1361\begin{bmatrix} 1.0000 & -.3461 \\ -.3461 & 1.0000 \end{bmatrix} = \begin{bmatrix} 1.1361 & -.3932 \\ -.3932 & 1.1361 \end{bmatrix}$$

Then we have

$$\underset{\mathbf{R}^{-1}}{\begin{bmatrix} 1.1361 & -.3932 \\ -.3932 & 1.1361 \end{bmatrix}} \underset{\mathbf{r}}{\begin{bmatrix} .2396 \\ .3756 \end{bmatrix}} \underset{=}{=} \underset{\hat{\mathbf{b}}}{\begin{bmatrix} .1245 \\ .3325 \end{bmatrix}}$$

where

$$\hat{b}_1 = (1.1361)(.2396) + (-.3932)(.3756) = .1245$$

and

$$\hat{b}_2 = (-.3932)(.2396) + (1.1361)(.3756) = .3325$$

In the preceding chapter we found that $s_Y = 2.7435$, $s_{X_1} = 3.0094$, and $s_{X_2} = 9.3045$. Then we also have

$$b_1 = \frac{s_Y}{s_{X_1}} \hat{b}_1 = \frac{2.7435}{3.0094}(.1245) = .1135$$

and

$$b_2 = \frac{s_Y}{s_{X_2}} \hat{b}_2 = \frac{2.7435}{9.3045}(.3325) = .0980$$

and these values of b_1 and b_2 are equal, within rounding errors, to those we obtained in the previous chapter.

5.4 A Multiple Regression Problem with Three *X* Variables

We now consider an example of multiple regression analysis with three X variables. The intercorrelations of the X variables and the correlations of the X variables with the Y variables were done by a computer. Various other calculations were done by a computer, but we will also illustrate some that can be done rather easily with a hand calculator.

We give the values of the X variables and the Y variable in Table 5.1 so that it will be possible for you to run the same example on a computer if one is available to you. Almost all, if not all, computer centers have available a standard package program that will do all of the necessary calculations for a multiple regression problem. Because both the SPSS* and BMDP† package programs are available at a large number of computer centers, we have used these programs for the analyses shown in this chapter. These programs are constantly being updated, but, in general, you should be able to relate the output you obtain with that given in this chapter.

---

* Nie, N. H., Hull, C. H., Jenkins, J. G., Steinbrenner, K., and Bent, D. H. *SPSS: Statistical Package for the Social Sciences* (2nd ed.). New York: McGraw-Hill, 1975.

† Dixon, W. J., and Brown, M. B. *BMDP-77 Biomedical Computer Programs, P-Series.* Berkeley: University of California Press, 1977.

TABLE 5.1 Values of X_1, X_2, X_3, and Y for a sample of 25 subjects

Subjects	X_1	X_2	X_3	Y
1	11	38	10	15
2	7	42	16	18
3	12	38	18	17
4	13	36	15	16
5	14	40	15	17
6	15	32	11	14
7	5	20	13	10
8	14	44	18	21
9	14	34	12	17
10	10	28	16	11
11	8	24	10	13
12	16	30	16	18
13	15	26	15	16
14	14	24	12	15
15	10	26	12	14
16	9	18	14	13
17	11	30	16	16
18	9	26	13	13
19	7	18	11	11
20	10	10	17	6
21	9	12	8	12
22	10	32	18	18
23	10	18	14	12
24	16	20	15	16
25	10	18	12	15

The X variables were entered into the regression equation sequentially, with X_1 being entered first, then X_2, and finally X_3. Table 5.2 gives the intercorrelations of the X variables and the correlations of the X variables with the Y variable. The means for X_1, X_2, X_3, and Y are given in the column at the right of Table 5.2, and the standard deviations are given at the bottom of the table.

TABLE 5.2 Intercorrelations of the X variables and correlations of the X variables with Y for the data in Table 5.1. Means are at the right and standard deviations at the bottom of the table.

	X_1	X_2	X_3	Y	Means
X_1	1.00000	.34606	.23961	.53722	11.16
X_2		1.00000	.37556	.77916	27.36
X_3			1.00000	.34608	13.88
Y				1.00000	14.56
s	3.00943	9.30448	2.74348	3.18957	

TABLE 5.3 Regression and residual sums of squares for the Y values in Table 5.1

Source of variation	Sum of squares	d.f.	Mean square	F
Regression	168.20529	3	56.06843	15.50183
Residual	75.95471	21	3.61689	
Total	244.16000	24		

5.5 Test of Significance of MS_{reg} and $R_{Y.123}$

Table 5.3 gives the computer output showing the partitioning of the total sum of squares into the sum of squares for regression and the residual sum of squares. The regression sum of squares has k degrees of freedom, where $k = 3$ is the number of X variables, and the residual sum of squares has $n - k - 1$ degrees of freedom. For the test of significance of the regression mean square we have $F = 15.50$, rounded, with 3 and 21 degrees of freedom; this is a significant value with $\alpha = .01$. The three X variables, as a set, account for a significant proportion of the total sum of squares.

The squared multiple correlation coefficient, $R^2_{Y.123}$, will be given by

$$R^2_{Y.123} = \frac{SS_{reg}}{SS_{tot}} = \frac{168.20529}{244.16000} = .68891$$

The F test for the significance of $R_{Y.123}$ will be given by

$$F = \frac{R^2_{Y.123}/k}{(1 - R^2_{Y.123})/(n - k - 1)} \tag{5.3}$$

and this F test is equivalent to the F test of MS_{reg}. For example, we have

$$F = \frac{.68891/3}{(1 - .68891)/21} = 15.5015$$

which is equal, within rounding errors, to the F test for MS_{reg}.

5.6 The Inverse of a Symmetric 3 × 3 Correlation Matrix

The inverse of a symmetric 3 × 3 correlation matrix, if it exists, can be calculated by finding $c\mathbf{T}$, where c is a constant or scalar and is equal to

$$c = \frac{1}{1 + (2)r_{12}r_{13}r_{23} - r^2_{12} - r^2_{13} - r^2_{23}}$$

and $\mathbf{T}$ is a symmetric 3×3 matrix given by

$$\mathbf{T} = \begin{bmatrix} 1 - r_{23}^2 & r_{13}r_{23} - r_{12} & r_{12}r_{23} - r_{13} \\ & 1 - r_{13}^2 & r_{12}r_{13} - r_{23} \\ & & 1 - r_{12}^2 \end{bmatrix}$$

The intercorrelations of the three X variables are given in Table 5.2. Substituting with these values, we obtain

$$c = \frac{1}{1 + (2)(.34606)(.23961)(.37556) - (.34606)^2 - (.23961)^2 - (.37556)^2}$$
$$= 1.34397$$

and

$$\mathbf{T} = \begin{bmatrix} .85895 & -.25607 & -.10964 \\ -.25607 & .94259 & -.29264 \\ -.10964 & -.29264 & .88024 \end{bmatrix}$$

Observe that in calculating the entries in $\mathbf{T}$ it is not necessary to calculate the entries below the diagonal, because the matrix is symmetric. Multiplying each of the entries in $\mathbf{T}$ by the scalar $c = 1.34397$, we obtain

$$\mathbf{R}^{-1} = \begin{bmatrix} 1.15440 & -.34415 & -.14735 \\ -.34415 & 1.26681 & -.39330 \\ -.14735 & -.39330 & 1.18302 \end{bmatrix}$$

5.7 Calculating the Regression Coefficients for Standardized Variables

The regression coefficients for the standardized variables will be given by $\mathbf{R}^{-1}\mathbf{r}$ and in our example we have

$$\underset{\mathbf{R}^{-1}}{\begin{bmatrix} 1.15440 & -.34415 & -.14735 \\ -.34415 & 1.26681 & -.39330 \\ -.14735 & -.39330 & 1.18302 \end{bmatrix}} \underset{\mathbf{r}}{\begin{bmatrix} .53722 \\ .77916 \\ .34608 \end{bmatrix}} = \underset{\hat{\mathbf{b}}}{\begin{bmatrix} .30102 \\ .66605 \\ .02382 \end{bmatrix}}$$

where, for example.

$$\hat{b}_1 = (1.15440)(.53722) + (-.34415)(.77916) + (-.14735)(.34608)$$
$$= .30102$$

$$\hat{b}_2 = (-.34415)(.53722) + (1.26681)(.77916) + (-.39330)(.34608)$$
$$= .66605$$

and

$$\hat{b}_3 = (-.14735)(.53722) + (-.39330)(.77916) + (1.18302)(.34608)$$
$$= .02382$$

5.8 Calculation of $R^2_{YY'}$ from the Values of $\hat{b}_i$ and r_{Yi}

With k variables in the regression equation it will also be true that

$$R^2_{YY'} = \hat{b}_1 r_{Y1} + \hat{b}_2 r_{Y2} + \hat{b}_3 r_{Y3} + \cdots + \hat{b}_k r_{Yk} \quad (5.4)$$

Substituting with the values of $\hat{b}_i$ that we have just calculated and with the corresponding correlation coefficients given in Table 5.2, we have

$$R^2_{Y.123} = (.30102)(.53722) + (.66605)(.77916) + (.02382)(.34608)$$
$$= .68892$$

which is equal, within rounding errors, to the value we obtained previously.

We should not, however, regard (5.4) as representing the unique contribution of each of the X variables to the regression sum of squares. The proportion of SS_{tot} accounted for by a given X variable depends on when it is entered into the regression equation and on what other X variables have preceded it.

5.9 Obtaining the Regression Coefficients for Unstandardized Variables

Recall that the values of the regression coefficients for unstandardized variables can be obtained from the values of $\hat{b}_i$ by means of the relationship

$$b_i = \frac{s_Y}{s_i} \hat{b}_i$$

where s_i is the standard deviation of X_i. The values of the standard deviations are given in Table 5.2, and thus we can obtain

$$b_1 = \frac{3.18957}{3.00943}(.30102) = .31904$$

$$b_2 = \frac{3.18957}{9.30448}(.66605) = .22832$$

and

$$b_3 = \frac{3.18957}{2.74348}(.02382) = .02769$$

5.10 The Standard Error of b_i

The standard error of a regression coefficient b_i when several X variables are involved will be given by

$$s_{b_i} = \sqrt{\frac{MS_{res}}{\Sigma x_i^2(1 - R_{i.}^2)}} \tag{5.5}$$

where $R_{i.}^2$ is the squared multiple correlation of X_i with the remaining X variables.* When the inverse of the correlation matrix between the X variables is available, the value of $R_{i.}^2$ is easily obtained. It will be given by

$$R_{i.}^2 = 1 - \frac{1}{r^{ii}} \tag{5.6}$$

where r^{ii} is the diagonal entry in the inverse of the correlation matrix for the ith X variable. In our example we have $r^{11} = 1.15440$, and

$$R_{1.}^2 = 1 - \frac{1}{1.15440} = .13375 = R_{1.23}^2$$

We have $MS_{res} = 3.61689$, $\Sigma x_1^2 = 217.36$, and $R_{1.23}^2 = .13375$. Then for the standard error of b_1 we have†

$$s_{b_1} = \sqrt{\frac{3.61689}{(217.36)(1 - .13375)}} = .13860$$

The standard errors of b_2 and b_3 can be obtained in the same manner in which we obtained the standard error of b_1.

Table 5.4 summarizes the calculations regarding the regression coefficients. The table gives the values of the regression coefficients $\hat{b}_i$ for standardized variables and also the values of b_i for the unstandardized variables. The standard errors for the values of b_i are shown along with the t and F tests for the values of b_i. The values of F are simply the values of t^2. Each of the t or F ratios in the table is based on 1 and 21 d.f. With $\alpha = .05$, b_1 and b_2 differ significantly from zero and b_3 does not.

---

* If all variables are in standardized form, then the standard error of $\hat{b}_i$ will be

$$s_{\hat{b}_i} = \sqrt{\frac{1 - R_{YY'}^2}{(n - k - 1)(1 - R_{i.}^2)}}$$

The t or F test of $\hat{b}_i$ will result in the same values as the t or F test of b_i, within rounding errors.

† We know that $\Sigma x_1^2 = (n - 1)s_1^2$. Values of the standard deviations are given in Table 5.2. For Σx_1^2 we have $(25 - 1)(3.00943)^2 = 217.36$. The other values of Σx_i^2 can be obtained in the same manner.

TABLE 5.4 Summary of calculations for the regression coefficients

Variable	$\hat{b}_i$	b_i	s_{b_i}	t	F
X_1	.30102	.31904	.13860	2.30188	5.29865
X_2	.66605	.22832	.04697	4.86097	23.62903
X_3	.02382	.02769	.15395	.17986	.03235

5.11 Calculating the Semipartial Correlation Coefficients

Although the F test for b_1 and the semipartial correlation coefficient $r_{Y(1.23)}$ are equivalent, as are the F tests for b_2 and $r_{Y(2.13)}$ and b_3 and $r_{Y(3.12)}$, it is often useful to know the value of the semipartial correlation coefficients for each X variable when it is entered last in the regression equation. Although computer programs often do not provide the values of the semipartial correlation coefficients, they can easily be obtained from the F (or t) test for the regression coefficients when all X variables are in the regression equation. For example, because it is true that

$$F_1 = \frac{b_1^2}{s_{b_1}^2} = \frac{r_{Y(1.23)}^2}{(1 - R_{Y.123}^2)/(n - k - 1)}$$

we also have

$$r_{Y(1.23)}^2 = F_1(1 - R_{Y.123}^2)/(n - k - 1)$$

where F_1 is the value of F for the test of significance of b_1. In our example we have

$$r_{Y(1.23)}^2 = 5.29865(1 - .68891)/21 = .07849$$

Then we know that when X_1 is entered last in the regression equation, it accounts for .07849 of the total sum of squares over and above that already accounted for by X_2 and X_3. This represents a significant increment in the regression sum of squares. The fact that $r_{Y(1.23)}^2$ is significant is equivalent to saying that $R_{Y.123}^2$ is significantly greater than $R_{Y.23}^2$.

In the same manner we can calculate

$$r_{Y(2.13)}^2 = (23.62903)(1 - .68891)/21 = .35004$$

Thus, we know that when X_2 is entered last in the regression equation, it accounts for .35004 of the total sum of squares beyond that already accounted for by X_1 and X_3. This also represents a significant increment in the regression sum of squares. We can also say that $R_{Y.123}^2$ is significantly larger than $R_{Y.13}^2$.

Similarly, we have

$$r^2_{Y(3.12)} = (.03235)(1 - .68891)/21 = .00048$$

Because b_3 is not significant, neither is $r_{Y(3.12)}$. Variable X_3 does not represent a significant increment in the regression sum of squares beyond that already accounted for by X_1 and X_2, and we can conclude that $R^2_{Y.123}$ is not significantly greater than $R^2_{Y.12}$.

The values of $r^2_{Y(1.23)}$=.07849, $r^2_{Y(2.13)}$=.35004, and $r^2_{Y(3.12)}$=.00048 give the proportion of the total sum of squares accounted for by each of the X variables when it is entered last in the regression equation. For example, when X_1 is entered last in the regression equation, it accounts for .07849 of the total sum of squares. When it is entered first in the regression equation, it accounts for $r^2_{Y1} = (.53722)^2 = .28861$ of the total sum of squares. Only X_1 and X_2 contribute significantly to the regression sum of squares when they are entered last in the regression equation.

5.12 Values of the Semipartial Correlation Coefficients Based on Sequential Runs

To check on the accuracy of the calculation of the semipartial correlation coefficients based on the tests of significance of the regression coefficients, three sequential runs were done on a computer with each X variable entered last in one of the runs. The results are shown in Table 5.5. Using the values given in Table 5.5, we have

$$r^2_{Y(1.23)} = R^2_{Y.123} - R^2_{Y.23} = .68891 - .61042 = .07849$$
$$r^2_{Y(2.31)} = R^2_{Y.123} - R^2_{Y.31} = .68891 - .33872 = .35019$$
$$r^2_{Y(3.12)} = R^2_{Y.123} - R^2_{Y.12} = .68891 - .68844 = .00047$$

The values we obtained are in agreement with those reported above within rounding errors.

TABLE 5.5 Results of three sequential runs with each X variable entered last in one of the runs

X_1 entered last		X_2 entered last		X_3 entered last	
X_2	.60709	X_3	.11977	X_1	.28861
X_2, X_3	.61042	X_3, X_1	.33872	X_1, X_2	.68844
X_2, X_3, X_1	.68891	X_3, X_1, X_2	.68891	X_1, X_2, X_3	.68891

5.13 Multicollinearity

The existence of high intercorrelations among the X variables is called the problem of *multicollinearity*. For example, if for any two X variables, X_i and X_j, we have $r_{ij} = 1.0$ or -1.0, then the inverse of the symmetric correlation matrix does not exist. Note, for example, that the scalar $1/(1 - r_{12}^2)$ used in obtaining the inverse of a 2×2 matrix becomes 1/0, an operation that is not permitted if r_{12} is equal to either 1.0 or -1.0. If r_{12}^2 approaches 1.0, then the scalar $1/(1 - r_{12}^2)$ becomes very large.

It is obviously not difficult to recognize a correlation between any two X variables that approaches -1.0 or 1.0, but in large correlation matrices it is possible for one of the X variables, say X_i, to be a linear function of the other remaining X variables such that $R_{i.}^2$, the squared multiple correlation of X_i with the remaining X variables, approaches 1.0. If $R_{i.}^2 = 1.0$, then the inverse of the correlation matrix does not exist. For example, suppose that for three X variables we have $r_{12} = .96$, $r_{13} = .80$, and $r_{23} = .60$. In this case the scalar c that multiplies the matrix $\mathbf{T}$ is equal to

$$c = \frac{1}{1 + (2)(r_{12})(r_{13})(r_{23}) - r_{12}^2 - r_{13}^2 - r_{23}^2}$$

and with the three values of r_{ij} just given we have $c = 1/0$, again an operation that is inadmissible. The inverse of this 3×3 correlation matrix does not exist. The reason it does not is that $R_{2.13}^2$ is equal to 1.0, although this may not be obvious from simply looking at the intercorrelation matrix.

Evidence regarding a high degree of multicollinearity may be provided by the standard errors of the regression coefficients. Note that when several X variables are involved, we have

$$s_{b_i} = \sqrt{\frac{MS_{res}}{\Sigma x_i^2(1 - R_{i.}^2)}}$$

where $R_{i.}^2$ is the squared multiple correlation of X_i with the remaining X variables. As $R_{i.}^2$ approaches 1, the denominator becomes small and s_{b_i} becomes large. An extremely large value of s_{b_i} for any of the regression coefficients may be an indication of multicollinearity.

When several X variables are highly correlated with each other and each in turn is correlated with Y, it is reasonable to believe that each of the X variables is accounting for the same common variance with Y. In this case, the semipartial correlations for each of these X variables, when entered last in the regression equation, may be quite low, even though each of the X variables may be substantially correlated with Y. For example, if X_i, X_j, and X_k are highly intercorrelated and each is correlated with Y, then when any one of the three is entered after the other two, it will probably not represent any substantial increase in the regression sum of squares.

If several X variables are highly intercorrelated, it may make sense to create a new X variable that is simply a composite of the several X variables. Another possible solution is to use only one of the X variables in the set of highly correlated X variables.*

Exercises

5.1 If **I** is a 3 × 3 identity matrix, what is the inverse of **I**?

5.2 Does the inverse of the following 3 × 3 matrix exist? Explain why or why not.

$$\begin{bmatrix} 1.00 & .96 & .80 \\ .96 & 1.00 & .60 \\ .80 & .60 & 1.00 \end{bmatrix}$$

5.3 What is the value of the multiple correlation coefficient $R_{2.13}$ in the matrix given in Exercise 5.2?

5.4 If the intercorrelations of five X variables are all equal to zero, what is the inverse of the correlation matrix?

5.5 If the intercorrelations of three X variables are all equal to zero, what will be the value of the semipartial correlation $r_{Y(1.23)}$? Explain why.

5.6 If four X variables are uncorrelated with each other, what will be the value of $R^2_{Y.1234}$?

5.7 If three X variables are uncorrelated with each other, what will be the values of the regression coefficients, $\hat{b}_i$, for the standardized X variables?

5.8 Suppose that with five X variables and one Y variable we have $MS_{res} = 1500.00$ and $\Sigma x_3^2 = 100.0$, and we find that $s_{b_3} = 27.4548$. What does this suggest? Explain why.

* Multicollinearity occurs primarily with nonexperimental data in which the X variables cannot be controlled. Various methods, other than ordinary least squares, have been proposed for estimating regression coefficients when multicollinearity exists among the X variables. For an application of one of these methods, *ridge regression*, to nonexperimental data, see Price (1977). A general discussion of these methods can be found in Darlington (1978).

5.9 If the squared multiple correlation of X_i with the other X variables in a multiple regression analysis is equal to 1.00, what will the standard error of the regression coefficient, b_i, be equal to? Explain your answer.

5.10 Prove that if all variables are in standardized form, then Equation (5.5) in the text reduces to

$$s_{\hat{b}_i} = \sqrt{\frac{1 - R_{YY'}^2}{(n - k - 1)(1 - R_{i.}^2)}}$$

5.11 The values of $\hat{b}_1$, $\hat{b}_2$, and $\hat{b}_3$ are given in Table 5.4. Calculate the standard error for each of these regression coefficients. The values of $R_{i.}^2$ can be obtained from the inverse matrix given in the chapter. Do the t tests of the standardized regression coefficients result in the same values, within rounding errors, as the t tests for b_1, b_2, and b_3?

5.12 In a footnote to Section 5.1, we pointed out that if all variables are in standardized form, then the regression equation becomes

$$Y' = \hat{b}_1 z_1 + \hat{b}_2 z_2 + \cdots + \hat{b}_k z_k$$

and that the variance of Y' will be equal to $R_{YY'}^2$.

(a) Show that the mean of the Y' values is equal to zero.

(b) In one of the examples described in the chapter we have $\hat{b}_1 = .1245$, $\hat{b}_2 = .3325$, $r_{12} = .3461$, and $R_{Y.12}^2 = .1547$. Show that if

$$Y' = .1245 z_1 + .3325 z_2$$

then $s_{Y'}^2 = .1547 = R_{Y.12}^2$.

(c) For the three-variable problem described in the chapter, the intercorrelations of the X variables are given in Table 5.2 and the values of $\hat{b}_i$ in Table 5.4. For this example show that if

$$Y' = \hat{b}_1 z_1 + \hat{b}_2 z_2 + \hat{b}_3 z_3$$

then $s_{Y'}^2 = .68891 = R_{Y.123}^2$.

5.13 Explain each of the following concepts:

identity matrix | multicollinearity

inverse of a matrix

6

Equivalence of the F Tests of the Null Hypotheses $\mu_1 - \mu_2 = 0$, $\rho = 0$, and $\beta = 0$

6.1 The t Test for the Difference Between $k = 2$ Independent Means

A commonly used experimental design is to assign n subjects at random to each of two treatments. In some cases one of the two treatments may be a placebo or a control. After the treatments have been applied, the means, $\bar{Y}_1$ and $\bar{Y}_2$, for the two groups on a dependent variable of interest are obtained, and the difference between the two means is divided by the standard error of the difference between the means to obtain a t ratio. Thus, we have

$$t = \frac{\bar{Y}_1 - \bar{Y}_2}{s_{\bar{Y}_1 - \bar{Y}_2}} \tag{6.1}$$

with 1 and $n_1 + n_2 - 2$ degrees of freedom. There is no need for n_1 to be equal to n_2, but we shall assume that they are. The results we present are perfectly general and applicable to the situation where $n_1 \neq n_2$.

Under the usual assumption that the Y values are independently and normally distributed with the same variance, $\sigma_{Y.1}^2 = \sigma_{Y.2}^2 = \sigma_Y^2$ for each of the two treatments, Treatment 1 and Treatment 2, the unbiased estimate of σ_Y^2 will be the mean square within treatments, or

$$MS_W = \frac{\Sigma(Y_1 - \bar{Y}_1)^2 + \Sigma(Y_2 - \bar{Y}_2)^2}{n_1 + n_2 - 2} = \frac{\Sigma y_1^2 + \Sigma y_2^2}{n_1 + n_2 - 2} \tag{6.2}$$

The values of $\bar{Y}_1$ and $\bar{Y}_2$ are assumed to be unbiased estimates of the corresponding population means, μ_1 and μ_2, and the null hypothesis to be tested is $\mu_1 - \mu_2 = 0$. The standard error of the difference between the two means will be given by

$$s_{\bar{Y}_1 - \bar{Y}_2} = \sqrt{MS_W\left(\frac{1}{n_1} + \frac{1}{n_2}\right)} \tag{6.3}$$

and if $n_1 = n_2 = n_i$, then

$$s_{\bar{Y}_1 - \bar{Y}_2} = \sqrt{\frac{2MS_W}{n_i}} \tag{6.4}$$

Having paid our respects to the standard t test for the difference between two means, we turn to the more useful equivalent $F = t^2$ and the analysis of variance.

6.2 An Orthogonal Partitioning of SS_{tot}

If we forget, for the moment, that the Y values can be classified into two groups corresponding to the two treatments, we can find the total sum of squares in the usual way. Thus,

$$SS_{tot} = \Sigma(Y - \bar{Y})^2$$

where $\bar{Y}$ is the mean of the $n_1 + n_2$ values of Y and the summation is over the complete set of $n_1 + n_2$ values. Then, taking into account that we have n_1 values of Y_1 associated with Treatment 1 and n_2 values of Y_2 associated with Treatment 2, we can find the means

$$\bar{Y}_1 = \frac{\Sigma Y_1}{n_1} \quad \text{and} \quad \bar{Y}_2 = \frac{\Sigma Y_2}{n_2}$$

and the sum of squared deviations within each treatment group about the mean of the treatment group, or

$$SS_W = \Sigma(Y_1 - Y_1)^2 + \Sigma(Y_2 - \bar{Y}_2) = \Sigma y_1^2 + \Sigma y_2^2 \tag{6.5}$$

If SS_W is subtracted from SS_{tot}, we have

$$SS_{tot} - SS_W = SS_T$$

where SS_T refers to the treatment of squares defined by

$$SS_T = n_1(\bar{Y}_1 - \bar{Y})^2 + n_2(\bar{Y}_2 - \bar{Y})^2 \tag{6.6}$$

and where $\bar{Y}_1$ and $\bar{Y}_2$ are the means for the two treatments and $\bar{Y}$ is the mean of all $n_1 + n_2$ observations. Thus, we see that SS_{tot} can be partitioned into the two components SS_W and SS_T.

To prove that SS_W and SS_T are *orthogonal*, we write the following identity:

$$Y - \bar{Y} = (\bar{Y}_i - \bar{Y}) + (Y - \bar{Y}_i) \tag{6.7}$$

When we square (6.7) and sum over the n_1 observations for Treatment 1, we have the product sum

$$2\Sigma(\bar{Y}_1 - \bar{Y})(Y - \bar{Y}_1)$$

But $\bar{Y}_1 - \bar{Y}$ is a constant for the n_1 observations in Treatment 1, and, consequently,

$$2(\bar{Y}_1 - \bar{Y})\Sigma(Y - \bar{Y}_1) = 0$$

because the sum of the deviations from the mean is equal to zero. Thus, the product sum for $(\bar{Y}_i - \bar{Y})(Y - \bar{Y}_i)$ will be equal to zero when we sum over the n_1 observations for Treatment 1 and also when we sum over the n_2 observations for Treatment 2. Thus, when (6.7) is squared and summed over all $n_1 + n_2$ observations, we have

$$\Sigma(Y - \bar{Y})^2 = \Sigma n_i(\bar{Y}_i - \bar{Y})^2 + \Sigma(Y - \bar{Y}_i)^2 \tag{6.8}$$

where $\Sigma(Y - \bar{Y})^2 = SS_{tot}$, or the sum of squared deviations of all Y values from the overall mean $\bar{Y}$; $\Sigma n_i(\bar{Y}_i - \bar{Y})^2 = n_1(\bar{Y}_1 - \bar{Y})^2 + n_2(\bar{Y}_2 - \bar{Y})^2 = SS_T$; and $\Sigma(Y - \bar{Y}_i)^2 = \Sigma(Y_1 - \bar{Y}_1)^2 + \Sigma(Y_2 - \bar{Y}_2)^2 = SS_W$. It is of some importance to note that (6.8) generalizes to the situation in which we have $k > 2$ treatment groups.

We let $n = n_1 + n_2 = 20$ or the total number of observations and $k = 2$ be the number of treatment groups. Then, as we have seen, we have partitioned SS_{tot} with $n - 1 = 19$ degrees of freedom into the two orthogonal components, SS_T with $k - 1 = 1$ d.f. and SS_W with $n - 2 = 18$ d.f. This partitioning of SS_{tot} in the analysis of variance bears a remarkable similarity to the partitioning of SS_{tot} in a linear regression analysis with one X variable. In the linear regression analysis, we partitioned SS_{tot} with $n - 1$ degrees of freedom into the two orthogonal components, SS_{reg} with 1 degree of freedom and SS_{res} with $n - 2$ degrees of freedom. The resemblance between these two partitionings of SS_{tot} is no accident. In fact, as we shall show later in the chapter,

$$SS_T = SS_{reg} \quad \text{and} \quad SS_W = SS_{res}$$

6.3 The F Test for the Difference Between $k = 2$ Independent Means

Table 6.1 gives the Y values for ten subjects tested under Treatment 1 and ten subjects tested under Treatment 2. Then the total sum of squares or the sum of squared deviations of the twenty Y values about the mean of all twenty Y values can be obtained from the algebraic identity

$$\Sigma(Y - \bar{Y})^2 = \Sigma Y^2 - \frac{(\Sigma Y)^2}{n_1 + n_2} = 806 - \frac{(120)^2}{20} = 86.00$$

Similarly, we have

$$\Sigma(Y_1 - \bar{Y}_1)^2 = 520 - \frac{(70)^2}{10} = 30.00$$

and

$$\Sigma(Y_2 - \bar{Y}_2)^2 = 286 - \frac{(50)^2}{10} = 36.00$$

and

$$SS_W = 30.00 + 36.00 = 66.00$$

with $n_1 + n_2 - 2$ degrees of freedom.

TABLE 6.1 Values of a dummy variable X and values of a dependent variable Y for subjects assigned at random to $k = 2$ treatments

Treatment 1			Treatment 2		
Subjects	X	Y	Subjects	X	Y
1	1	10	11	0	7
2	1	6	12	0	5
3	1	5	13	0	3
4	1	10	14	0	8
5	1	7	15	0	7
6	1	6	16	0	4
7	1	7	17	0	6
8	1	8	18	0	5
9	1	6	19	0	3
10	1	5	20	0	2
Σ	10	70	Σ	0	50

The treatment sum of squares can be obtained from the algebraic identity

$$SS_T = n_1(\bar{Y}_1 - \bar{Y})^2 + n_2(\bar{Y}_2 - \bar{Y})^2 = \frac{(\Sigma Y_1)^2}{n_1} + \frac{(\Sigma Y_2)^2}{n_2} - \frac{(\Sigma Y)^2}{n_1 + n_2} \quad (6.9)$$

and for the data in Table 6.1 we have

$$SS_T = \frac{(70)^2}{10} + \frac{(50)^2}{10} - \frac{(120)^2}{20} = 20.00$$

with degrees of freedom equal to $k - 1$, where k is equal to the number of treatment groups.

Then the F test of the null hypothesis $\mu_1 - \mu_2 = 0$ will be given by

$$F = \frac{SS_T/(k-1)}{SS_W/(n_1 + n_2 - 2)} = \frac{MS_T}{MS_W} \quad (6.10)$$

In the present example we have

$$F = \frac{20.00/1}{66.00/18} = 5.455$$

with 1 and 18 d.f; this is a significant value of F with $\alpha = .05$.

6.4 Dummy Coding for an Experiment Involving $k = 2$ Treatments

In a linear regression analysis with one X variable, the values of the independent variable X were assumed to be fixed and represented ordered values of some quantitative variable. In the experiment described the independent variable X simply consists of the different treatments. The two treatments are fixed in the sense that any conclusions drawn from the results of the experiment are assumed to apply only to the two treatments investigated.

It is obvious that we can differentiate the two treatments by an X variable that consists of any two different numbers.* For example, we could let $X = 1$ represent Treatment 1 and $X = 2$ represent Treatment 2; or we could let $X = 1$ represent Treatment 1 and $X = -1$ represent Treatment 2. Because the values of X do not necessarily represent ordered values of a quantitative variable, we shall refer to the values

* The correlation coefficient between a variable X that has only two possible values, such as 1 and 0 or -1 and 1, and a variable Y that has many possible values is called a *point biserial correlation coefficient.*

See notes

↑ This is important ↑

of X as an X vector. The X vector simply serves to identify or differentiate between the two treatments without implying that the values in the X vector represent quantitative ordered values of a variable. In the present example we shall choose $X = 1$ to represent Treatment 1 and $X = 0$ to represent Treatment 2. The use of $X = 1$ and $X = 0$ to identify treatment groups is commonly referred to as *dummy coding.*

6.5 A Regression Analysis of the Data in Table 6.1

With dummy coding we will have n_1 values of $X = 1$ and n_2 values of $X = 0$, where n_1 and n_2 are the number of observations for Treatments 1 and 2, respectively. For the data in Table 6.1 we have $n = n_1 + n_2$ and

$$\Sigma(X - \bar{X})^2 = \Sigma X^2 - \frac{(\Sigma X)^2}{n} = n_1 - \frac{(n_1)^2}{n} = \frac{n_1 n_2}{n}$$

or, with $n_1 = n_2 = 10$,

$$\Sigma(X - \bar{X})^2 = \frac{(10)(10)}{20} = 5$$

For the product sum of the X and Y values in Table 6.1 we have

$$\Sigma(X - \bar{X})(Y - \bar{Y}) = \Sigma XY - \frac{(\Sigma X)(\Sigma Y)}{n} = 70 - \frac{(10)(120)}{20} = 10$$

Note that ΣXY will be equal to ΣY_1 or the treatment sum for the observations identified by $X = 1$.* Then for the regression coefficient we have

$$b = \frac{\Sigma xy}{\Sigma x^2} = \frac{10}{5} = 2.0$$

and we note that

$$b = \bar{Y}_1 - \bar{Y}_2 = 7.0 - 5.0 = 2.0$$

or the difference between the two treatment means.†

We have $\Sigma X = 10$ and $\bar{X} = 10/20 = .5$. Then with $\bar{Y} = 6.0$ we have

$$a = \bar{Y} - b\bar{X} = 6.0 - (2.0)(.5) = 5.0$$

---

* With a single X vector and dummy coding, ΣXY will always be equal to ΣY_1 or the treatment sum for those observations identified by $X = 1$.

† With dummy coding and a single X vector, b, will always be equal to the difference between the two treatment means, $\bar{Y}_1 - \bar{Y}_2$.

which is simply the mean of the treatment group identified by $X = 0$.*
For the regression equation we have

$$Y' = a + bX = 5.0 + 2.0X$$

For each of the n_1 values of $X = 1$ we see that

$$Y' = 5.0 + (2.0)(1) = 7.0 = \bar{Y}_1$$

and for each of the n_2 values of $X = 0$,

$$Y' = 5.0 + (2.0)(0) = 5.0 = \bar{Y}_2$$

Then for the residual sum of squares we have

$$SS_{res} = \Sigma(Y - Y')^2 = \Sigma(Y_1 - \bar{Y}_1)^2 + \Sigma(Y_2 - \bar{Y}_2)^2 = SS_W$$

with $n_1 + n_2 - 2 = 18$ d.f.

Recall that the regression sum of squares is equal to $\Sigma(Y' - \bar{Y})^2$. But we have n_1 values of $Y' = \bar{Y}_1$ and n_2 values of $Y' = \bar{Y}_2$. Then the regression sum of squares will be

$$SS_{reg} = \Sigma(Y' - \bar{Y})^2 = n_1(\bar{Y}_1 - \bar{Y})^2 + n_2(\bar{Y}_2 - \bar{Y})^2 = SS_T$$

with 1 d.f. We also know that $SS_{reg} = (\Sigma xy)^2/\Sigma x^2$. We have $\Sigma xy = 10$ and $\Sigma x^2 = 5$, and we see that $SS_{reg} = (10)^2/5 = 20$, which is also equal to SS_T.

For the F test of the significance of MS_{reg} we have

$$F = \frac{SS_{reg}/(k - 1)}{SS_{res}/[n - (k - 1) - 1]} = \frac{MS_{reg}}{MS_{res}} = \frac{MS_T}{MS_W} \tag{6.11}$$

where $k = 2$ is the number of treatment groups and $k - 1 = 1$ is the number of X vectors. We see that the F test of the regression mean square in a linear regression analysis is equivalent to the F test of the treatment mean square in an analysis of variance of the same data. Both tests involve exactly the same distribution assumptions.

6.6 F Test of the Null Hypotheses $\rho = 0$ and $\beta = 0$

With a single X variable we have previously shown that the regression sum of squares divided by the total sum of squares was equal to a squared correlation coefficient. Then, because $SS_{reg} = SS_T$, we have

$$r^2 = \frac{SS_{reg}}{SS_{tot}} = \frac{SS_T}{SS_{tot}} \tag{6.12}$$

* With dummy coding and a single X vector, the value of a will always be equal to the treatment mean for those observations identified by $X = 0$, or $\bar{Y}_2$ in our example.

For the data in Table 6.1 we have

$$r^2 = \frac{20.0}{86.0} = .232558$$

We can also calculate

$$r^2 = \frac{(\Sigma xy)^2}{(\Sigma x^2)(\Sigma y^2)} = \frac{(10)^2}{(5)(86)} = .232558 \tag{6.13}$$

It is easy to see that r^2 as defined by (6.12) and (6.13) must be equal, because $SS_{reg} = (\Sigma xy)^2/\Sigma x^2$ and $SS_{tot} = \Sigma y^2$, and

$$r^2 = \frac{SS_{reg}}{SS_{tot}} = \frac{(\Sigma xy)^2/\Sigma x^2}{\Sigma y^2} = \frac{(\Sigma xy)^2}{(\Sigma x^2)(\Sigma y^2)}$$

Note also that $r^2 SS_{tot} = SS_{reg} = SS_T$ and $(1 - r^2)SS_{tot} = SS_{res} = SS_W$; r^2 is simply the proportion of the total sum of squares accounted for by X or the two treatments, and $1 - r^2$ is the proportion of SS_{tot} that is independent of the two treatments.

For the F test of the null hypothesis $\rho = 0$ we have

$$F = t^2 = \frac{r^2}{1 - r^2}(n - 2)$$

and multiplying both numerator and denominator by SS_{tot} we have

$$F = \frac{r^2 SS_{tot}}{(1 - r^2)SS_{tot}}(n - 2) = \frac{MS_{reg}}{MS_{res}} = \frac{MS_T}{MS_W}$$

For the F test of the null hypothesis $\beta = 0$ we have

$$F = t^2 = \frac{b^2}{s_b^2} = \frac{(\Sigma xy)^2/(\Sigma x^2)^2}{MS_{res}/\Sigma x^2} = \frac{(\Sigma xy)^2/\Sigma x^2}{MS_{res}} = \frac{MS_{reg}}{MS_{res}} = \frac{MS_T}{MS_W}$$

6.7 Summary

We have shown that in the analysis of variance, SS_{tot} can be partitioned into the two orthogonal components SS_W and SS_T. We also showed that in a linear regression analysis of the same data, SS_{tot} can be partitioned into the two components SS_{res} and SS_{reg}, and that $SS_{res} = SS_W$ and $SS_{reg} = SS_T$. The F test of MS_T in the analysis of variance and the F test of MS_{reg} in linear regression are equivalent. We also showed that the F test of the null hypothesis $\rho = 0$ and the F test of the null hypothesis $\beta = 0$ are equivalent to each other and to the F test of the null hypothesis $\mu_1 - \mu_2 = 0$.

Exercises

6.1 In an experiment, twenty subjects were divided at random into two groups of ten subjects each. Subjects in one group were administered a low dosage of a drug, and subjects in the other group were administered a placebo. Performance of both groups of subjects was then measured on a dependent variable to determine whether the drug had any influence. Measures for the subjects receiving the drug (Treatment 1) and the placebo (Treatment 2) are shown below:

Treatment 1 (Drug)	Treatment 2 (Placebo)
8	5
10	4
10	4
9	2
6	3
7	6
7	6
9	5
6	2
8	3

(a) Calculate t, and determine whether the means for Treatment 1 and Treatment 2 differ significantly.

(b) Calculate $F = MS_T/MS_W$, and show that F is equal to the value of t^2.

(c) Use dummy coding, assigning $X = 1$ to those subjects receiving Treatment 1 and $X = 0$ to those subjects receiving Treatment 2. Calculate the correlation coefficient, r, between X and Y. Show that the value of r^2 is equal to SS_T/SS_{tot}.

(d) In the regression equation $Y' = a + bX$, is the value of a equal to the mean for Treatment 1 or the mean for Treatment 2?

(e) In the regression equation $Y' = a + bX$, the values of a and b are related to values of the treatment means. What are these relationships?

(f) Calculate $SS_{reg} = (\Sigma xy)^2/\Sigma x^2$. Is this value equal to SS_T in the analysis of variance?

(g) Is $F = MS_{reg}/MS_{res} = MS_T/MS_W$?

(h) In the regression equation $Y' = a + bX$, what is the value of Y' when $X = 0$? What is the value of Y' when $X = 1$?

(i) Test the regression coefficient and the correlation coefficient for significance, using the t test. Show that the values of t^2 are equal to $F = MS_T/MS_W$.

6.2 Suppose that we have an experiment involving two treatments but with unequal n's. A simple example is given below.

Treatment 1	Treatment 2
6	3
4	5
5	4
2	1
3	2
7	
8	

(a) Calculate the t test and the F test for the difference between the two treatment means. Is t^2 equal to F?

(b) Use dummy coding, assigning $X = 1$ to subjects in Treatment 1 and $X = 0$ to subjects in Treatment 2. Calculate r, and test it for significance using the t test. Is the value of t obtained equal to the value of t for the test of the difference between the two means?

(c) Calculate the values of a and b in the regression equation $Y' = a + bX$. Is the value of a equal to the mean for Treatment 2? Is the value of b equal to the difference between the means for Treatment 1 and Treatment 2?

(d) Test b for significance, using the t test. Is the value of t obtained equal to the value of t for the test of significance of the difference between the two means?

(e) Is $SS_{reg} = (\Sigma xy)^2/\Sigma x^2$ equal to SS_T? Is SS_{res} equal to SS_W?

6.3 Explain each of the following concepts:

dummy coding

treatment sum of squares

mean square within treatments

7

Completely Randomized Designs with One Treatment Factor

7.1 The Analysis of Variance

In Table 7.1 we give the outcome of an experiment in which five subjects were assigned at random to each of three treatments. The values of Y, the dependent variable, were obtained after the application of the treatments. It is not necessary to have an equal number of subjects assigned to each treatment. The same principles of analysis apply to the situation in which the n's for the various treatments are not equal.

TABLE 7.1 Values of Y for an experiment with $k = 3$ treatments

	T_1	T_2	T_3
	5	1	10
	12	2	13
	9	7	16
	8	3	12
	11	8	17
Σ	45	21	68

For the analysis of variance of the data in Table 7.1, we first calculate

$$SS_{tot} = (5)^2 + (12)^2 + \cdots + (17)^2 - \frac{(134)^2}{15} = 322.93$$

with $15 - 1 = 14$ d.f. The sums for the three treatments are 45, 21, and 68, respectively. Then for the treatment sum of squares we have

$$SS_T = \frac{(45)^2}{5} + \frac{(21)^2}{5} + \frac{(68)^2}{5} - \frac{(134)^2}{15} = 220.93$$

with 2 d.f., and by subtraction we obtain

$$SS_W = 322.93 - 220.93 = 102.00$$

with 12 d.f.

Table 7.2 summarizes the analysis of variance. As a test of the null hypothesis $\mu_1 = \mu_2 = \mu_3$, we have $F = MS_T/MS_W = 12.996$ with 2 and 12 d.f., a significant value with $\alpha = .05$.

TABLE 7.2 Summary of the analysis of variance for the data in Table 7.1

Source of variation	Sum of squares	d.f.	Mean square	F
Treatments	220.93	2	110.465	12.996
Within treatments	102.00	12	8.500	
Total	322.93	14		

7.2 Dummy Coding

Now let us consider a multiple regression analysis of the same experimental data. For any experiment of the kind described, with k treatments, $k - 1$ X vectors will jointly serve to identify the treatment groups to which the subjects have been assigned. There are many ways in which these X vectors may be constructed. One simple way is by dummy coding, using 1's and 0's. With this method of coding, all members of the first treatment group are assigned 1 in X_1 and all other subjects are assigned 0. For the second vector, X_2, the subjects in Treatment 2 are assigned 1 and all others are assigned 0. Note that these two X vectors serve jointly to identify the subjects in each of the $k = 3$ treatment groups. For example, subjects assigned to Treatment 1 are identified by (1, 0), subjects assigned to Treatment 2 are identified by (0, 1), and subjects assigned to Treatment 3 are identified by (0, 0), as shown in Table 7.3.

If we had $k = 4$ treatment or groups, then dummy coding with three X vectors would identify the members in each treatment group. In this instance the subjects assigned to Treatment 1 would be identified by (1, 0, 0), the subjects assigned to Treatment 2 would be identified by (0, 1, 0), the subjects assigned to Treatment 3 would be identified by (0, 0, 1) and the subjects assigned to Treatment 4 would be identified by (0, 0, 0).

TABLE 7.3 Dummy coding for the data in Table 7.1

	Subjects	X_1	X_2	Y
Treatment 1	1	1	0	5
	2	1	0	12
	3	1	0	9
	4	1	0	8
	5	1	1	11
Treatment 2	6	0	1	1
	7	0	1	2
	8	0	1	7
	9	0	1	3
	10	0	1	8
Treatment 3	11	0	0	10
	12	0	0	13
	13	0	0	16
	14	0	0	12
	15	0	0	17

7.3 The Multiple Regression Equation with Dummy Coding

The multiple regression equation with two X vectors will be

$$Y' = a + b_1X_1 + b_2X_2$$

With dummy coding for X_1 and X_2, the least squares estimates* of a, b_1, and b_2 can be shown to be $a = \bar{Y}_3$, $b_1 = \bar{Y}_1 - \bar{Y}_3$, and $b_2 = \bar{Y}_2 - \bar{Y}_3$. Then we have

$$Y' = \bar{Y}_3 + (\bar{Y}_1 - \bar{Y}_3)X_1 + (\bar{Y}_2 - \bar{Y}_3)X_2$$

For the data in Table 7.3 we have $\bar{Y}_3 = 13.6$, $\bar{Y}_1 = 9.0$, and $\bar{Y}_2 = 4.2$. Consequently, we have

$$Y' = 13.6 + (9.0 - 13.6)X_1 + (4.2 - 13.6)X_2$$

We now note that for each of the subjects assigned to Treatment 1 we have

$$Y' = 13.6 + (9.0 - 13.6)(1) + (4.2 - 13.6)(0) = 9.0 = \bar{Y}_1$$

Similarly, for each of the subjects assigned to Treatment 2 we have

$$Y' = 13.6 + (9.0 - 13.6)(0) + (4.2 - 13.6)(1) = 4.2 = \bar{Y}_2$$

and for each of the subjects assigned to Treatment 3 we have

$$Y' = 13.6 + (9.0 - 13.6)(0) + (4.2 - 13.6)(0) = 13.6 = \bar{Y}_3$$

Consequently, for the residual sum of squares we have

$$\begin{aligned}\Sigma(Y - Y')^2 &= \Sigma[Y - (a + b_1X_1 + b_2X_2)]^2 \\ &= \Sigma(Y_1 - \bar{Y}_1)^2 + \Sigma(Y_2 - \bar{Y}_2)^2 + \Sigma(Y_3 - \bar{Y}_3)^2\end{aligned}$$

where the summation on the left is over all $n_1 + n_2 + n_3$ observations and the summations on the right are over the n_1, n_2, and n_3 observations, respectively, for the three treatment groups. We see that the residual sum of squares in the regression analysis is equal to the within treatment sum of squares in the analysis of variance.

Recall also that the regression sum of squares is equal to $\Sigma(Y' - \bar{Y})^2$. We will have n_1 values of $Y' = \bar{Y}_1$, n_2 values of $Y' = \bar{Y}_2$, and n_3 values of $Y' = \bar{Y}_3$. Thus, we have

$$SS_{reg} = \Sigma(Y' - \bar{Y})^2 = n_1(\bar{Y}_1 - \bar{Y})^2 + n_2(\bar{Y}_2 - \bar{Y})^2 + n_3(\bar{Y}_3 - \bar{Y})^2 = SS_T$$

* With dummy coding, the value of a will always be equal to the mean for those observations identified by 0's in all X vectors or the observations for the kth treatment. We will have $k - 1$ X vectors, and the values for $b_1, b_2, \ldots, b_{k-1}$ will always be equal to the difference between the means, $\bar{Y}_1 - \bar{Y}_k$, $\bar{Y}_2 - \bar{Y}_k, \ldots, \bar{Y}_{k-1} - \bar{Y}_k$, respectively.

We also know that the regression sum of squares will be given by

$$SS_{reg} = b_1\Sigma x_1 y + b_2\Sigma x_2 y + \cdots + b_k\Sigma x_k y$$

and for the data in Table 7.3 we have

$$\Sigma x_1 y = \Sigma Y_1 - \frac{n_1 \Sigma Y}{\Sigma n_i} = 45 - \frac{(5)(134)}{15} = .3333$$

where Σn_i is the total number of observations. Similarly, we have

$$\Sigma x_2 y = \Sigma Y_2 - \frac{n_2 \Sigma Y}{\Sigma n_i} = 21 - \frac{(5)(134)}{15} = -23.6667$$

In our example we have $b_1 = \bar{Y}_1 - \bar{Y}_3 = 9.0 - 13.6 = -4.6$ and $b_2 = \bar{Y}_2 - \bar{Y}_3 = 4.2 - 13.6 = -9.4$. Then we also have

$$SS_{reg} = (-4.6)(.3333) + (-9.4)(-23.6667) = 220.93$$

which, as we now know, must be equal to $SS_T = 220.93$.

7.4 The Squared Multiple Correlation Coefficient

Regardless of the nature or number of the X vectors or variables, the squared multiple correlation coefficient between Y and Y' will be given by

$$R^2_{YY'} = 1 - \frac{\Sigma(Y - Y')^2}{SS_{tot}} = \frac{SS_{tot} - SS_{res}}{SS_{tot}} = \frac{SS_{reg}}{SS_{tot}}$$

With dummy coding for the $k = 3$ treatments we have $SS_{reg} = SS_T$, and for the data in Table 7.3 we have

$$R^2_{Y.12} = \frac{220.93}{322.93} = .6841$$

7.5 Test of Significance of $R_{YY'}$

In the analysis of variance, $F = MS_T/MS_W$ is, as we have pointed out, a test of the null hypothesis $\mu_1 = \mu_2 = \mu_3 = \cdots = \mu_k$. In the multiple regression analysis, the equivalent test is $F = MS_{reg}/MS_{res}$, or, in terms of $R_{YY'}$,

$$F = \frac{R^2_{YY'}/(k-1)}{(1 - R^2_{YY'})/[n - (k-1) - 1]} \tag{7.1}$$

where k is the number of treatment groups and $k - 1$ is the number of X vectors used to define the treatment groups. In our example we have

$$F = \frac{.6841/2}{(1 - .6841)/12} = 12.99$$

The reason the F ratio defined by (7.1) is equal to $F = MS_T/MS_W$ is that

$$R^2_{YY'} = \frac{SS_{reg}}{SS_{tot}} = \frac{SS_T}{SS_{tot}}$$

and, consequently,

$$SS_T = SS_{reg} = R^2_{YY'}SS_{tot}$$

and

$$\begin{aligned} SS_W = SS_{res} &= SS_{tot} - SS_T \\ &= SS_{tot} - R^2_{YY'}SS_{tot} \\ &= (1 - R^2_{YY'})SS_{tot} \end{aligned}$$

Multiplying both the numerator and denominator of (7.1) by SS_{tot}, we see that the F test of the multiple correlation coefficient is equal to $F = MS_T/MS_W$.

In Section 4.7 we stated that the F test of the multiple correlation coefficient was a test of the null hypothesis that in the population $\beta_1 = \beta_2 = \cdots = \beta_k = 0$. In our example, $b_1 = \bar{Y}_1 - \bar{Y}_3$ is the sample estimate of $\beta_1 = \mu_1 - \mu_3$, and $b_2 = \bar{Y}_2 - \bar{Y}_3$ is the sample estimate of $\beta_2 = \mu_2 - \mu_3$. If the null hypothesis $\beta_1 = \beta_2 = 0$ is true, then the null hypothesis $\mu_1 = \mu_2 = \mu_3$ will also have to be true.

7.6 Multiple Regression Analysis with Nonorthogonal and Orthogonal *X* Variables

As we pointed out in Section 4.5, when the X variables in a multiple regression analysis are correlated, there is no satisfactory method of partitioning the regression sum of squares so as to be able to determine the unique contribution of each of the X variables. The proportion of the total sum of squares accounted for by a given X variable will depend on when it is entered into the regression equation and which other X variables have preceded it. This is not true, however, if the X variables are mutually orthogonal.

If the X variables are mutually orthogonal, then

$$R^2_{Y.123\ldots k} = r^2_{Y1} + r^2_{Y2} + r^2_{Y3} + \cdots + r^2_{Yk}$$

In this case the contribution of each of the X variables to the value of $R^2_{YY'}$ and to the regression sum of squares is unique and independent of the contribution of every other X variable. No matter in what order or in what combination the X variables are entered into the regression equation, the values of r^2_{Yi} will give the proportion of the total sum of squares, $\Sigma(Y - \bar{Y})^2$, that can be accounted for by X_i. It will also be true that with uncorrelated X variables, the regression coefficient $b_i = \Sigma x_i y / \Sigma x_i^2$ and the regression sum of squares, $SS_{reg_i} = (\Sigma x_i y)^2 / \Sigma x_i^2$ for any variable X_i, will remain exactly the same no matter in which order or in what combination the X variables are entered into the regression equation. Then the total regression sum of squares can be partitioned into a set of orthogonal components such that

$$SS_{reg} = \frac{(\Sigma x_1 y)^2}{\Sigma x_1^2} + \frac{(\Sigma x_2 y)^2}{\Sigma x_2^2} + \cdots + \frac{(\Sigma x_k y)^2}{\Sigma x_k^2}$$

where the regression sums of squares on the right correspond to each of the mutually orthogonal X variables.

7.7 $R^2_{YY'}$ with Orthogonal Coding

Table 7.4 repeats the Y values given in Table 7.3 but with two new X vectors. The important thing to note about these two X vectors is that $\Sigma X_1 = \Sigma X_2 = 0$, so that $\Sigma x_1^2 = \Sigma X_1^2$ and $\Sigma x_2^2 = \Sigma X_2^2$ and the product sum $\Sigma(X_1 - \bar{X}_1)(X_2 - \bar{X}_2) = \Sigma X_1 X_2 = 0$. The two X vectors are, therefore, orthogonal. With two orthogonal X vectors we have

$$R^2_{Y.12} = r^2_{Y1} + r^2_{Y2}$$

TABLE 7.4 Orthogonal coding for the data in Table 7.1

	Subjects	X_1	X_2	Y
	1	2	0	5
	2	2	0	12
Treatment 1	3	2	0	9
	4	2	0	8
	5	2	0	11
	6	−1	1	1
	7	−1	1	2
Treatment 2	8	−1	1	7
	9	−1	1	3
	10	−1	1	8
	11	−1	−1	10
	12	−1	−1	13
Treatment 3	13	−1	−1	16
	14	−1	−1	12
	15	−1	−1	17

For the data in Table 7.4 it is easy to calculate

$$r_{Y1}^2 = \frac{(\Sigma x_1 y)^2}{(\Sigma x_1^2)(\Sigma y^2)} = \frac{(1)^2}{(30)(322.93)} = .0001$$

and

$$r_{Y2}^2 = \frac{(\Sigma x_2 y)^2}{(\Sigma x_2^2)(\Sigma y^2)} = \frac{(-47)^2}{(10)(322.93)} = .6840$$

Then we have

$$R_{Y.12}^2 = .0001 + .6840 = .6841$$

as before.

7.8 Orthogonal Comparisons or Contrasts

The two X vectors in Table 7.4 correspond to what, in the analysis of variance, are called two *orthogonal comparisons* or *contrasts.** Note that because $\Sigma X_1 = 0$, we have $\Sigma x_1 y = \Sigma X_1 Y$, and, in Table 7.4, this is simply $\Sigma X_1 Y = 2\Sigma Y_1 - (\Sigma Y_2 + \Sigma Y_3)$. Because $n_i \bar{Y}_i = \Sigma Y_i$, this comparison can also be written as†

$$2n_i \bar{Y}_1 - (n_i \bar{Y}_2 + n_i \bar{Y}_3) = n_i[2\bar{Y}_1 - (\bar{Y}_2 + \bar{Y}_3)]$$

because we have an equal number of observations for each treatment. The population value for the comparison represented by X_1 will be

$$n_i[2\mu_1 - (\mu_2 + \mu_3)]$$

and the null hypothesis to be tested is that $2\mu_1 - (\mu_2 + \mu_3) = 0$.

The regression sum of squares for the comparison represented by X_1 will be equal to

$$SS_{reg_1} = b_1 \Sigma x_1 y = \frac{(\Sigma x_1 y)^2}{\Sigma x_1^2} = \frac{(\Sigma X_1 Y)}{\Sigma X_1^2}$$

with 1 d.f. Using the values given in Table 7.4, we have

$$\Sigma X_1 Y = 2\Sigma Y_1 - (\Sigma Y_2 + \Sigma Y_3) = (2)(45) - (21 + 68) = 1$$

---

* With equal n's for each treatment, X_i will be a comparison or contrast if $\Sigma X_i = 0$, and two comparisons, X_i and X_j, will be orthogonal if $\Sigma X_i X_j = 0$. With unequal n's, X_i will be a comparison or contrast if $\Sigma n_i X_i = 0$, and two comparisons, X_i and X_j, will be orthogonal if $\Sigma n_i X_i X_j = 0$.

† The nature of the comparison represented by X_1 remains unchanged by multiplication or division of the values in X_1 by a constant. We could, for example, multiply each of the values 2, -1, and -1 in X_1 by $1/2n_i$, where n_i is the number of observations for each treatment. This would result in a new X_1 vector with values of $1/n_i$, $-1/2n_i$, and $-1/2n_i$. Then we would have the comparison $\Sigma X_1 Y = \bar{Y}_1 - \frac{1}{2}(\bar{Y}_2 + \bar{Y}_3)$, and the null hypothesis corresponding to this comparison would be $\mu_1 - \frac{1}{2}(\mu_2 + \mu_3) = 0$ instead of $n_i[2\mu_1 - (\mu_2 + \mu_3)] = 0$. The two null hypotheses are obviously equivalent.

We also have $\Sigma X_1^2 = 30$, and the regression sum of squares will be equal to $(1)^2/30 = .03333$. For the residual sum of squares we have

$$SS_{res} = (1 - R_{Y.12}^2)SS_{tot} = (1 - .6841)(322.93) = 102.01358$$

with $n - (k - 1) - 1$ degrees of freedom, where $k - 1$ is the total number of X vectors.

For the test of significance of the comparison represented by X_1 we have

$$F = \frac{.03333}{102.01358/12} = .0039$$

which is not a significant value of F with $\alpha = .05$ and with 1 and 12 d.f. Equivalently, we have the F test for r_{Y1}, or

$$F = \frac{r_{Y1}^2}{(1 - .6841)/12} = \frac{.0001}{(1 - .6841)/12} = .0038$$

which is equal, within rounding errors, to the F obtained in the test of significance of the regression mean square.

For the second X vector in Table 7.4 we have the comparison*

$$\Sigma X_2 Y = \Sigma Y_2 - \Sigma Y_3 = 21 - 68 = -47$$

and the null hypothesis to be tested is that in the population

$$n_i\mu_2 - n_i\mu_3 = n_i(\mu_2 - \mu_3) = 0$$

or that $\mu_2 - \mu_3 = 0$. The regression sum of squares for this comparison is equal to

$$SS_{reg_2} = b_2\Sigma x_2 y = \frac{(\Sigma x_2 y)^2}{\Sigma x_2^2} = \frac{(\Sigma X_2 Y)^2}{\Sigma X_2^2} = \frac{(-47)^2}{10} = 220.90$$

and we observe that

$$SS_{reg} = SS_{reg_1} + SS_{reg_2} = .0333 + 220.9000 = 220.93$$

rounded, which is equal to SS_T.

Then we have

$$F = \frac{MS_{reg2}}{MS_{res}} = \frac{220.90}{102.01358/12} = 25.985$$

We have also calculated $r_{Y2}^2 = .6840$, and for the F test of this correlation coefficient we have

$$F = \frac{.6840}{(1 - .6841)/12} = 25.983$$

* Again, note that we could have coded the values in X_2 as $0/n_i$, $1/n_i$, and $-1/n_i$. Then the comparison represented by X_2 would be $\bar{Y}_2 - \bar{Y}_3$, and the null hypothesis would be $\mu_2 - \mu_3 = 0$. This null hypothesis is obviously equivalent to the null hypothesis $n_i(\mu_2 - \mu_3) = 0$.

which is equal, within rounding errors, to $F = MS_{reg_2}/MS_{res}$. The tests of significance of r_{Y2} and MS_{reg_2} are equivalent, and both are tests of the null hypothesis $n_i(\mu_2 - \mu_3) = 0$.

7.9 Orthogonal Partitioning of SS_T and SS_{reg}

If we have a completely randomized design with k treatments and with an equal number of observations for each treatment, it is always possible to partition the treatment sum of squares in the analysis of variance into $k - 1$ orthogonal comparisons or contrasts, each with 1 d.f. and such that

$$SS_T = SS_1 + SS_2 + \cdots + SS_{k-1}$$

In a multiple regression analysis of the same data, each of the comparisons or contrasts in the analysis of variance can be represented by an X vector. There will be a total of $k - 1$ such X vectors, one for each comparison, and they will be mutually orthogonal. Consequently, the regression sum of squares can also be partitioned into a set of orthogonal components, each with 1 d.f. and such that

$$SS_{reg} = SS_{reg_1} + SS_{reg_2} + \cdots + SS_{reg_{k-1}}$$

For every orthogonal partitioning of SS_T in the analysis of variance there is a corresponding orthogonal partitioning of SS_{reg} in a multiple regression analysis.

7.10 Orthogonal Coding for $k > 3$ Treatments with Equal n's

With equal n's for each treatment group, orthogonal X vectors are easily constructed. Table 7.4 shows the coding for $k = 3$ treatments. With $k = 4$ treatments we would have

	Values of X		
	X_1	X_2	X_3
Treatment 1	3	0	0
Treatment 2	−1	2	0
Treatment 3	−1	−1	1
Treatment 4	−1	−1	−1

The pattern is the same for $k > 4$ treatments.

We emphasize that there are many other ways in which mutually orthogonal X vectors can be constructed and that some sets may be more meaningful than others in a given experiment. In the set shown above, the test of significance of r_{Y1} corresponds to a test of the null hypothesis $\mu_1 - \frac{1}{3}(\mu_2 + \mu_3 + \mu_4) = 0$; the test of significance of r_{Y2} corresponds to a test of the null hypothesis $\mu_2 - \frac{1}{2}(\mu_3 + \mu_4) = 0$; and a test of significance of r_{Y3} corresponds to a test of the null hypothesis $\mu_3 - \mu_4 = 0$. Because these X vectors are mutually orthogonal, it will also be true that

$$R^2_{Y.123} = r^2_{Y1} + r^2_{Y2} + r^2_{Y3}$$

and

$$SS_{tot}R^2_{Y.123} = SS_{tot}(r^2_{Y1} + r^2_{Y2} + r^2_{Y3}) = SS_{reg}$$

or

$$SS_{reg} = SS_{reg_1} + SS_{reg_2} + SS_{reg_3}$$

And SS_{reg} will be equal to SS_T in the analysis of variance. It will also be true that

$$SS_{tot}(1 - R^2_{Y.123}) = SS_{res}$$

and SS_{res} will be equal to SS_W in the analysis of variance.

7.11 Equivalent Tests of Significance in a Multiple Regression Analysis

With orthogonal X vectors, the F tests of significance of r_{Yi}, b_i, and MS_{reg_i} are all equivalent. It simply doesn't matter which statistic we test for significance; each of the three tests is a test of the same null hypothesis, and each will result in the same F ratio, within rounding errors. We have some preference for the tests of significance of the correlation coefficients, because, having calculated r_{Yi}, we have the additional information provided by r^2_{Yi}, that is, the proportion of the total sum of squares accounted for by a given comparison or contrast. A comparison or contrast may be statistically significant with $\alpha = .05$, but an experimenter should be concerned not only with statistical significance but also with "practical" significance. If $r_{Yi} = .10$ is statistically significant, we can then ask whether $r^2_{Yi} = (.10)^2 = .01$ represents any practical contribution to the regression sum of squares.

good

Exercises

7.1 We have $k = 4$ treatment groups, and we use dummy coding to identify the observations in each treatment group.

(a) How many X vectors will be needed? Explain why.

(b) In the regression equation $Y' = a + b_1X_1 + b_2X_2 + b_3X_3$, what will the values of a, b_1, b_2, and b_3 be equal to if we use dummy coding?

7.2 Suppose we have an equal number of observations in each treatment group. We let n_i be the number in each treatment group, and $n = \Sigma n_i$. Prove that with dummy coding the correlation coefficient between any two of the X vectors will be $r_{ij} = -1/(k-1)$, where k is the number of treatment groups.

7.3 If we have $k = 4$ treatments and identify the observations in each treatment group by means of orthogonal X vectors, what will the value of $R^2_{Y.123}$ be equal to?

7.4 The outcome of an experiment involving $k = 4$ treatment is given below:

Treatment 1	Treatment 2	Treatment 3	Treatment 4
8	14	6	3
6	11	7	5
7	10	9	2
5	12	10	4
4	13	8	6

(a) What are the values of SS_{tot}, SS_W, SS_T, and $F = MS_T/MS_W$?
(b) Use dummy coding to calculate $R^2_{Y.123}$.
(c) Show that $R^2_{Y.123}$ is equal to SS_T/SS_{tot}.
(d) Show that $SS_{tot}(1 - R^2_{Y.123})$ is equal to SS_W and that $SS_{tot}R^2_{Y.123}$ is equal to SS_T.
(e) Calculate the values of b_1, b_2, and b_3.
(f) Show that $b_1\Sigma x_1 y + b_2\Sigma x_2 y + b_3\Sigma x_3 y$ is equal to SS_T.

7.5 For the example given in Exercise 7.4, use orthogonal coding. Calculate

$$R^2_{Y.123} = r^2_{Y1} + r^2_{Y2} + r^2_{Y3}$$

7.6 If an orthogonal partitioning is made of the treatment sum of squares in the analysis of variance, can a similar orthogonal partitioning be made of the regression sum of squares in a multiple regression analysis of the same data? Explain why or why not.

7.7 With $k = 4$ treatments and with $n = 5$ subjects assigned at random to each treatment, we have $SS_{tot} = 250.0$. The experimenter uses coding with the X vectors shown below:

	X_1	X_2	X_3	Treatment Sums
Treatment 1	1	1	1	$\Sigma Y_1 = 10$
Treatment 2	1	−1	−1	$\Sigma Y_2 = 20$
Treatment 3	−1	1	−1	$\Sigma Y_3 = 30$
Treatment 4	−1	−1	1	$\Sigma Y_4 = 50$

(a) Are these X vectors orthogonal? Explain why or why not.
(b) What is the value of $R^2_{Y.123}$?
(c) What is the null hypothesis tested by X_1?
(d) What is the null hypothesis tested by X_2?
(e) What is the null hypothesis tested by X_3?
(f) What are the values of SS_T, SS_W, and $F = MS_T/MS_W$?
(g) Calculate $b_1\Sigma x_1 y$, $b_2\Sigma x_2 y$, and $b_3\Sigma x_3 y$. Is the sum of these three values equal to SS_T? Explain why or why not.
(h) Test r_{Y1}, r_{Y2}, and r_{Y3} for significance using an F test.
(i) With orthogonal X vectors, the standard error for b_i will be

$$s_{b_i} = \sqrt{\frac{MS_{res}}{\Sigma x_i^2}}$$

Calculate the standard error for each regression coefficient, and test each coefficient for significance using a t test. Are the squares of these values of t equal to the F tests r_{Yi} in (h)?
(j) The values of $b_i\Sigma x_i y$ in (g) give an orthogonal partitioning of SS_{reg}. Test each of these regression mean squares for significance using an F test. Are these values of F equal to those obtained in (h)?

7.8 Explain each of the following concepts:

dummy coding

orthogonal comparisons or contrasts

orthogonal partitioning of SS_T or SS_{reg}

orthogonal X variables or vectors

8

A Completely Randomized Design with a Quantitative Treatment Factor

8.1 Introduction

The treatments in a completely randomized experiment may consist of equally spaced values of a quantitative X variable. In experiments of this kind we are interested not only in the differences in the treatment means but also in determining if the treatment means are functionally related to the ordered values of the X variable. In this chapter we

TABLE 8.1 Outcome of an experiment with four equally spaced dosages of a drug

	T_1	T_2	T_3	T_4
	10	9	14	17
	8	13	13	15
	12	12	11	14
	11	10	12	18
	9	11	15	16
ΣY_i	50	55	65	80
$\bar{Y}_i$	10	11	13	16

examine the analysis of variance and multiple regression analysis of an experiment in which the treatments consist of four dosages of a drug such that X_1 is the smallest dose, $X_2 = 2X_1$, $X_3 = 3X_1$, and $X_4 = 4X_1$. The dependent variable Y is a physiological measure that presumably is influenced by the amount of the drug administered. Table 8.1 gives the outcome of the experiment in which $n_i = 5$ subjects were assigned at random to each of the four dosages of the drug.

8.2 Analysis of Variance for the Experiment

For the analysis of variance of the data in Table 8.1 we calculate the total sum of squares, or

$$SS_{tot} = (10)^2 + (8)^2 + \cdots + (16)^2 - \frac{(250)^2}{20} = 145.0$$

and the treatment sum of squares, or

$$SS_T = \frac{(50)^2}{5} + \frac{(55)^2}{5} + \frac{(65)^2}{5} + \frac{(80)^2}{5} - \frac{(250)^2}{20} = 105.0$$

Then, by subtraction, we obtain the sum of squares within treatments, or

$$SS_W = 145.0 - 105.0 = 40.0$$

The summary of the analysis of variance is shown in Table 8.2. For the test of significance of the treatment mean square we have $F = 14.0$, which is a significant value with $\alpha = .05$ and with 3 and 16 d.f.

TABLE 8.2 Summary of the analysis of variance for the data in Table 8.1

Source of variation	Sum of squares	d.f.	Mean square	F
Treatments	105.0	3	35.0	14.0
Within treatments	40.0	16	2.5	
Total	145.0	19		

8.3 Trend of the Y Means

Figure 8.1 shows a plot of the Y means against the values of X; it is obvious that the Y means increase with an increase in X. There is also a slight curvature in the trend of the Y means. If all of the Y means fell on a straight line with slope $b \neq 0$, then we know that all of the variation in the Y means could be accounted for by an equation of the first degree or by a linear equation. That part of SS_T that can be accounted for by an equation of the first degree is referred to as the linear component of SS_T, or SS_L, with 1 d.f.

We know that all of the variation in the Y means can be accounted for by an equation of degree no higher than $k - 1$, where k is the number of X values or the number of treatments, because any set of k points can be fitted perfectly by an equation of degree equal to $k - 1$. For example, a set of $k = 4$ means can be perfectly predicted by

$$\bar{Y}_i' = a + b_1X_1 + b_2X_1^2 + b_3X_1^3$$

We will make use of this fact in Section 8.5.

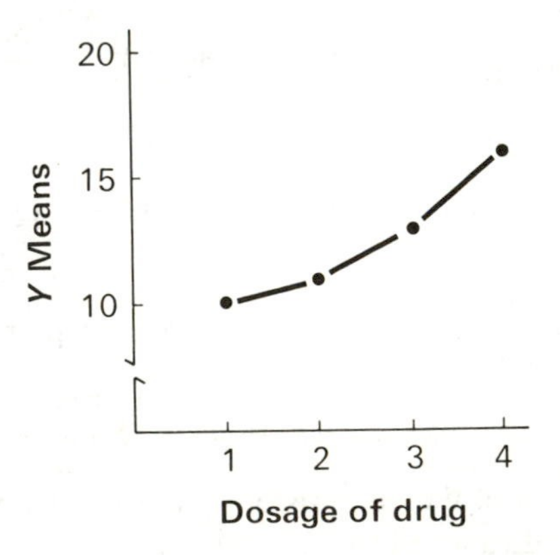

Figure 8.1 Plot of the Y means in Table 8.1 for each dosage of a drug.

TABLE 8.3 Values of $\bar{Y}_i$ and coefficients for orthogonal polynomials X_1, X_2, and X_3 for $k = 4$ equally spaced values of a quantitative variable

	T_1	T_2	T_3	T_4
$\bar{Y}_i$	10	11	13	16
X_1	−3	−1	1	3
X_2	1	−1	−1	1
X_3	−1	3	−3	1

8.4 Coefficients for Orthogonal Polynomials

Instead of working with polynomial equations of the second degree or higher, we make use of coefficients for orthogonal polynomials.* These coefficients enable us to investigate systematically the trend of a set of k means to find out which components of the trend are significant using only an equation of the first degree. Table III in the appendix gives coefficients for orthogonal polynomials for $k = 3$ to $k = 10$ values of X. Table 8.3 gives these coefficients for $k = 4$ values of X. Note that there are only $k - 1 = 3$ sets of orthogonal coefficients corresponding to the linear, quadratic, and cubic components, and that for each set of coefficients we have

$$\Sigma X_1 = \Sigma X_2 = \Sigma X_3 = 0$$

The sets of coefficients are also mutually orthogonal because for any two sets, X_i and X_j, we have

$$\Sigma(X_i - \bar{X}_i)(X_j - \bar{X}_j) = \Sigma X_i X_j = 0$$

* These coefficients are applicable in those experiments in which the values of X are equally spaced and in which we have an equal number of observations for each value of X. Coefficients for orthogonal polynomials can be derived when the values of X are not equally spaced or when the n_i's are unequal, but it would be impractical to table these coefficients for all possible cases. See J. Gaito, Unequal intervals and unequal n's in trend analysis, *Psychological Bulletin,* 1965, 63, 125–127, for methods that can be used to derive the coefficients for unequal n's or for values of X that are not equally spaced.

8.5 The Regression Equation Using Coefficients for Orthogonal Polynomials

We have said that any set of k means can be perfectly predicted by an equation of degree no higher than $k - 1$. With coefficients for orthogonal polynomials, in other words, the linear equation with $k = 4$ means

$$\bar{Y}'_i = \bar{Y} + b_1X_1 + b_2X_2 + b_3X_3 \tag{8.1}$$

will result in a set of predicted values equal to the corresponding observed values of the means. The successive terms b_1X_1, b_2X_2, and b_3X_3 correspond to terms of the first, second, and third degree in a polynomial equation. Not all of these terms may be needed to describe accurately the trend of a set of $k = 4$ means.

In (8.1) the values of X_1, X_2, and X_3 are known and can be obtained from Table III in the appendix. The corresponding values of b_1, b_2, and b_3 are unknown, but they can easily be obtained from the data of the experiment. For each set of X coefficients we have $\Sigma X_i = 0$. Then we also have

$$\Sigma(X_i - \bar{X}_i)(\bar{Y}_i - \bar{Y}) = \sum X_i(\bar{Y}_i - \bar{Y}) = \Sigma X_i\bar{Y}_i$$

and

$$\Sigma(X_i - \bar{X}_i)^2 = \Sigma X_i^2$$

For $\Sigma X_1\bar{Y}_i$ we have

$$\Sigma X_1\bar{Y}_i = (-3)(10) + (-1)(11) + (1)(13) + (3)(16) = 20.0$$

Similarly, we obtain

$$\Sigma X_2\bar{Y}_i = (1)(10) + (-1)(11) + (-1)(13) + (1)(16) = 2.0$$

and

$$\Sigma X_3\bar{Y}_i = (-1)(10) + (3)(11) + (-3)(13) + (1)(16) = 0$$

and because $\Sigma X_3\bar{Y}_i = 0$, b_3 must be equal to zero.

For b_1 we have

$$b_1 = \frac{\Sigma X_1\bar{Y}_i}{\Sigma X_1^2} = \frac{20}{20} = 1.0$$

and for b_2 we have

$$b_2 = \frac{\Sigma X_2\bar{Y}_i}{\Sigma X_2^2} = \frac{2}{4} = .5$$

For the overall mean we have $\bar{Y} = 12.5$, and the regression equation for predicting the Y means will be

$$\bar{Y}'_i = 12.5 + 1.0X_1 + .5X_2$$

For $\bar{Y}_1$ we have

$$\bar{Y}'_1 = 12.5 + (1.0)(-3) + (.5)(1) = 10.0$$

and this is exactly equal to the observed value $\bar{Y}_1 = 10$. Similarly, for the other three Y means we have

$$\bar{Y}'_2 = 12.5 + (1)(-1) + (.5)(-1) = 11.0$$
$$\bar{Y}'_3 = 12.5 + (1)(1) + (.5)(-1) = 13.0$$
$$\bar{Y}'_4 = 12.5 + (1)(3) + (.5)(1) = 16.0$$

8.6 Partitioning the Treatment Sum of Squares

We now want to find out how much of SS_T, the treatment sum of squares, can be accounted for by the linear component or by an equation of the first degree. The squared correlation coefficient between $\bar{Y}_i$ and X_1 will be given by

$$r^2_{\bar{Y}_i 1} = \frac{(\Sigma X_1 \bar{Y}_i)^2}{(\Sigma X_1^2)[\Sigma(\bar{Y}_i - \bar{Y})^2]} \tag{8.2}$$

Because $\Sigma(\bar{Y}_i - \bar{Y})^2 = SS_T/n_i = 105/5 = 21$, we have

$$r^2_{\bar{Y}_i 1} = \frac{(20)^2}{(20)(21)} = .95238$$

Similarly, we have

$$r^2_{\bar{Y}_i 2} = \frac{(2)^2}{(4)(21)} = .04762$$

Because X_1 and X_2 are orthogonal, we have

$$R^2_{\bar{Y}_i.12} = r^2_{\bar{Y}_i 1} + r^2_{\bar{Y}_i 2} = .95238 + .04762 = 1.00$$

and all of the variation in the Y means can be accounted for by the regression equation

$$\bar{Y}'_i = \bar{Y} + b_1 X_1 + b_2 X_2$$

The linear component of SS_T will be given by

$$SS_T r^2_{\bar{Y}_i 1} = (105)(.95238) = 100 = SS_L$$

rounded, and the quadratic component will be given by

$$SS_T r^2_{\bar{Y}_i 2} = (105)(.04762) = 5 = SS_Q$$

rounded. Note that $SS_L + SS_Q = SS_T$. For the test of significance of the linear component we have

$$F = \frac{MS_L}{MS_W} = \frac{100}{2.5} = 40.0$$

a significant value with $\alpha = .05$ and with 1 and 16 d.f. Testing the quadratic component for significance, we have

$$F = \frac{MS_Q}{MS_W} = \frac{5}{2.5} = 2.0$$

a nonsignificant value with $\alpha = .05$.

8.7 Multiple Regression Analysis of the Experiment Using Coefficients for Orthogonal Polynomials

Table 8.4 repeats the values of Y for each treatment group and shows the three X vectors based on coefficients for orthogonal polynomials. Observe that the three X vectors are mutually orthogonal. Because $\Sigma X_1 = 0$, we have $\Sigma x_1 y = \Sigma X_1 Y$ and

$$\begin{aligned}\Sigma X_1 Y &= (-3)(\Sigma Y_1) + (-1)(\Sigma Y_2) + (1)(\Sigma Y_3) + (3)(\Sigma Y_4)\\ &= (-3)(50) + (-1)(55) + (1)(65) + (3)(80)\\ &= 100\end{aligned}$$

TABLE 8.4 Coding of the data in Table 8.1 with coefficients for orthogonal polynomials

X_1	X_2	X_3	Y
−3	1	−1	10
−3	1	−1	8
−3	1	−1	12
−3	1	−1	11
−3	1	−1	9
−1	−1	3	9
−1	−1	3	13
−1	−1	3	12
−1	−1	3	10
−1	−1	3	11
1	−1	−3	14
1	−1	−3	13
1	−1	−3	11
1	−1	−3	12
1	−1	−3	15
3	1	1	17
3	1	1	15
3	1	1	14
3	1	1	18
3	1	1	16

and

$$\Sigma x_1^2 = \Sigma X_1^2 = 5[(-3)^2 + (-1)^2 + (1)^2 + (3)^2] = 100$$

Then the squared correlation coefficient between Y and X_1 will be

$$r_{Y1}^2 = \frac{(100)^2}{(100)(145)} = .68966$$

rounded.

Similarly, we have

$$\Sigma x_2 y = \Sigma X_2 Y = (1)(50) + (-1)(55) + (-1)(65) + (1)(80) = 10$$

$$\Sigma x_2^2 = \Sigma X_2^2 = 5[(1)^2 + (-1)^2 + (-1)^2 + (1)^2] = 20$$

Then for the squared correlation coefficient between Y and X_2 we have

$$r_{Y2}^2 = \frac{(10)^2}{(20)(145)} = .03448$$

Because X_1 and X_2 are orthogonal, we also have

$$R_{Y.12}^2 = r_{Y1}^2 + r_{Y2}^2 = .68966 + .03448 = .72414$$

8.8 Tests of Significance with a Multiple Regression Analysis

The tests of significance

$$F = \frac{r_{Y1}^2}{(1 - R_{Y.123}^2)/[n - (k - 1) - 1]}$$

and

$$F = \frac{r_{Y2}^2}{(1 - R_{Y.123}^2)/[n - (k - 1) - 1]}$$

where $k - 1 = 3$ is the number of X vectors, will be equivalent to the F tests of MS_L and MS_Q in the analysis of variance.* Substituting in the above expressions with $r_{Y1}^2 = .68966$, $r_{Y2}^2 = .03448$, and $R_{Y.123}^2 = .72414$, we have

$$F = \frac{.68966}{(1 - .72414)/16} = 40.0$$

* Even though X_3 does not contribute to the regression sum of squares, we presumably did not know this prior to the outcome of the experiment. Just as we would have used $(1 - R_{Y.123}^2)/[n - (k - 1) - 1]$ as the denominator in the F tests if $R_{Y.123}^2$ has been larger than $R_{Y.12}^2$, we continue to use $1 - R_{Y.123}^2$ with $[n - (k - 1) - 1]$ degrees of freedom even though $R_{Y.123}^2 = R_{Y.12}^2$. To do so is in agreement with the usual F tests in the analysis of variance.

rounded, and

$$F = \frac{.03448}{(1 - .72414)/16} = 2.0$$

rounded, and these two values of F are equal to those we obtained in the analysis of variance when we tested MS_L and MS_Q for significance.

With $R^2_{Y.123}$ equal to .72414, the regression sum of squares will be

$$SS_{reg} = SS_{tot}R^2_{Y.123} = (145)(.72414) = 105$$

which is equal to SS_T in the analysis of variance. Because X_1 and X_2 are orthogonal, SS_{reg} can be partitioned into the two orthogonal components $SS_{tot}r^2_{Y1}$ and $SS_{tot}r^2_{Y2}$. Thus, we have

$$SS_{reg} = (145)(.68966) + (145)(.03448) = 100 + 5$$

and these two values are equal to the linear and quadratic components of SS_T in the analysis of variance.

For the residual sum of squares we have

$$SS_{res} = SS_{tot}(1 - R^2_{Y.123}) = (145)(1 - .72414) = 40$$

which is equal to SS_W in the analysis of variance. The tests of significance of the two orthogonal components, 100 and 5, of the regression sum of squares will result in the same F ratios that we obtained in testing r_{Y1} and r_{Y2} for significance.

Note that the two regression coefficients,

$$b_1 = \frac{\Sigma X_1 Y}{\Sigma X_1^2} = \frac{100}{100} = 1.0$$

and

$$b_2 = \frac{\Sigma X_2 Y}{\Sigma X_2^2} = \frac{10}{20} = .5$$

are equal to the values we obtained in the analysis of variance. For the tests of significance of b_1 and b_2 we have

$$F = \frac{b_1^2}{MS_{res}/\Sigma X_1^2} = \frac{(1.0)^2}{(40/16)/100} = 40.0$$

and

$$F = \frac{b_2^2}{MS_{res}/\Sigma X_2^2} = \frac{(.5)^2}{(40/16)/20} = 2.0$$

which are also equal to the F tests of r_{Y1} and r_{Y2}.

TABLE 8.5 Coding of the data in Table 8.4 with powered X vectors

X_1	X_2	X_3	Y
1	1	1	10
1	1	1	8
1	1	1	12
1	1	1	11
1	1	1	9
2	4	8	9
2	4	8	13
2	4	8	12
2	4	8	10
2	4	8	11
3	9	27	14
3	9	27	13
3	9	27	11
3	9	27	12
3	9	27	15
4	16	64	17
4	16	64	15
4	16	64	14
4	16	64	18
4	16	64	16

8.9 Multiple Regression Analysis of the Experiment Using Powered X Vectors

In Table 8.5 we repeat the Y values in Table 8.4 but with three new X vectors. The values in X_1 correspond to the dosages of the drug, 1, 2, 3, and 4. For the second X vector we have $X_2 = X_1^2$, and for the third vector we have $X_3 = X_1^3$. In Table 8.6 we show the intercorrelations of the X vectors and their correlations with Y. As we might expect, these X vectors are highly correlated* with each other and also with the values of Y.

We are interested in a sequential analysis in which we first find out the proportion of the total sum of squares accounted for by the linear

* This matrix has a high degree of multicollinearity. If any one of the three X variables is entered first in the regression equation, it will account for a large proportion of the total sum of squares. The sequential analysis we have used assigns priority to the linear (X_1), followed by the quadratic (X_2) and cubic (X_3), components and is often referred to as a *hierarchical analysis*. In a hierarchical analysis, multicollinearity is not a problem because our interest is only in the increment in the regression sum of squares based on a prior and logical ordering of the variables. In other words, it is the semipartial correlations that are of interest.

TABLE 8.6 Intercorrelations of the X vectors in Table 8.5 and correlations of the X vectors with the values of Y

	X_1	X_2	X_3	Y	Means
X_1	1.00000	.98437	.95137	.83045	12.5
X_2		1.00000	.99053	.85018	2.5
X_3			1.00000	.84705	7.5
Y				1.00000	25.0
s	1.1471	5.8264	25.0788	2.7625	

component, or X_1, the additional increment attributable to the quadratic component, X_2, and finally whether the cubic component contributes to the regression sum of squares beyond that already accounted for by the linear and quadratic. We know that the cubic component will not, because we have already done the same analysis using coefficients for orthogonal polynomials. The present analysis is simply to show that the analyses using coefficients for orthogonal polynomials and powered X vectors are comparable.

With the sequential analysis we have

$$R^2_{Y.123} = r^2_{Y1} + r^2_{Y(2.1)} + r^2_{Y(3.12)}$$

and the terms on the right will give the proportion of the total sum of squares accounted for by the linear component, the additional increment attributable to the quadratic component, and the increment attributable to the cubic component over that already accounted for by the linear and quadratic components. Entering the X vectors in the regression equation sequentially, we have

$$R^2_{Y.1} = .68966$$
$$R^2_{Y.12} = .72414$$
$$R^2_{Y.123} = .72414$$

We see that the linear component $r^2_{Y1} = R^2_{Y.1}$ accounts for .68966 of the total sum of squares. For the squared semipartial correlation we have $r^2_{Y(2.1)} = R^2_{Y.12} - R^2_{Y.1} = .72414 - .68966 = .03448$; this is the proportion of the total sum of squares accounted for by the quadratic component after taking into account the linear component. Note that these two values are exactly the same as those we obtained using coefficients for orthogonal polynomials. The cubic component contributes nothing to the regression sum of squares, which, of course, was to be expected.

8.10 Tests of Significance with Powered X Vectors

The tests of significance

$$F = \frac{R^2_{Y.1}}{(1 - R^2_{Y.123})/[n - (k - 1) - 1]} = \frac{.68966}{(1 - .72414)/16} = 40.0$$

and

$$F = \frac{R^2_{Y.12} - R^2_{Y.1}}{(1 - R^2_{Y.123})/[n - (k - 1) - 1]} = \frac{.03448}{(1 - .72414)/16} = 2.0$$

result in the same F ratios that we obtained using coefficients for orthogonal polynomials.

Exercises

8.1 In an experiment involving five equally spaced values of a quantitative variable X, $n_i = 5$ subjects were assigned at random to each value of X. The values of Y, the dependent variable, are given below for each of the subjects for each value of X.

	X Variable				
	1	2	3	4	5
	10	12	9	14	15
	8	14	13	13	13
	12	11	12	11	12
	11	10	10	12	16
	9	13	11	15	14
Σ	50	60	55	65	70

(a) Calculate SS_T, SS_W, and $F = MS_T/MS_W$.

(b) Using coefficients for orthogonal polynomials, partition SS_T into the linear, quadratic, cubic, and quartic components. The quartic component of SS_T will be that part of SS_T accounted for by b_4X_4 in the regression equation shown in (c).

(c) Calculate the regression coefficients b_1, b_2, b_3, and b_4, and use the regression equation

$$\bar{Y}'_i = \bar{Y} + b_1X_1 + b_2X_2 + b_3X_3 + b_4X_4$$

to predict the Y means.

(d) Test the linear, quadratic, cubic, and quartic components of SS_T for significance.

(e) Calculate the correlation coefficient between the Y means and the coefficients X_1 for the linear component. Calculate each of the other correlation coefficients also. Show that

$$R^2_{\bar{Y}.1234} = r^2_{\bar{Y}_i 1} + r^2_{\bar{Y}_i 2} + r^2_{\bar{Y}_i 3} + r^2_{\bar{Y}_i 4} = 1.0$$

8.2 For the experiment described in Exercise 8.1, use a multiple regression analysis with coefficients for orthogonal polynomials as the X vectors.

(a) Is $F = MS_{reg}/MS_{res}$ equal to $F = MS_T/MS_W$ in the analysis of variance?

(b) Test r_{Y1}, r_{Y2}, r_{Y3}, and r_{Y4} for significance. Are these F ratios equal to the corresponding F ratios in the analysis of variance?

(c) Will the F test of significance for b_1 result in the same value of F as the F test of significance for r_{Y1}?

(d) Will the F test for MS_{reg_L} result in the same value of F as the F test for b_1 and the F test for r_{Y1}?

8.3 If you have access to a computer, use the original values for X_1, that is, 1, 2, 3, 4, and 5, to obtain a new X_1 vector where the values are $X - 3$. The new X_1 vector will have values of -2, -1, 0, 1, and 2. Then let $X_2 = X_1^2$, $X_3 = X_1^3$, and $X_4 = X_1^4$. Enter the X variables sequentially in the regression equation, that is, in the order X_1, X_2, X_3, and X_4. Then you can easily calculate the squared semipartial correlation coefficients shown on the right below:

$$R^2_{Y.1234} = r^2_{Y1} + r^2_{Y(2.1)} + r^2_{Y(3.12)} + r^2_{Y(4.123)}$$

The tests of significance of the semipartial correlation coefficients will result in the same F values as the tests of significance using coefficients for orthogonal polynomials in the analysis of variance or in the multiple regression analysis.

8.4 Explain each of the following concepts:

coefficients for orthogonal polynomials

cubic component

hierarchical analysis

linear component

powered X variables or vectors

quadratic component

quartic component

9

Factorial Experiments

9.1 The Nature of a Factorial Experiment with Two Factors

In some experiments, two or more independent variables are included. The independent variables are commonly referred to as *factors*, and the number of ways in which the variables are varied are commonly referred to as the number of *levels* of the factors. For example, in a word recognition experiment we might have the same list of words printed in two different typefaces. The typeface would be a factor A, with two levels, A_1 and A_2, corresponding to the two different typefaces. A second factor might consist of length of time of exposure of the words; this might be varied in two ways, with one exposure time of 45 milliseconds and another of 90 milliseconds. This factor would be factor B, with two levels, B_1 and B_2, corresponding to the two exposure times. Then a treatment would consist of a combination of

TABLE 9.1 Outcome of a 2 × 2 factorial experiment

	A_1B_1	A_1B_2	A_2B_1	A_2B_2
	7	5	3	11
	8	6	4	12
	9	7	5	13
	10	8	6	14
	11	9	7	15
Σ	45	35	25	65

one level from each factor, and we would have a total of $2 \times 2 = 4$ treatment combinations, if all are to be included in the experiment. They would be A_1B_1, A_1B_2, A_2B_1, and A_2B_2.

An experiment of this type is called a *factorial experiment*. If an equal number of subjects are assigned at random to each of the treatment combinations, the experimental design will be a balanced orthogonal design.* What this means is that not only can the total sum of squares be partitioned into the two orthogonal components SS_T and SS_W, but the treatment sum of squares itself can be partitioned into a set of meaningful orthogonal components: a sum of squares for the A factor (SS_A), with $a - 1$ degrees of freedom, where a is the number of levels of A; a sum of squares for the B factor (SS_B), with $b - 1$ degrees of freedom, where b is the number of levels of B; and a sum of squares for the $A \times B$ interaction (SS_{AB}), with $(a - 1)(b - 1)$ degrees of freedom. We shall see how this is done by considering the simple example shown in Table 9.1, where we have two levels of A and two levels of B, with $n = 5$ subjects assigned at random to each of the A_iB_j treatment combinations.

9.2 Analysis of Variance of the Data in Table 9.1

We proceed in the usual way by first calculating the total sum of squares and then the treatment sum of squares. We have

$$SS_{tot} = (7)^2 + (8)^2 + \cdots + (15)^2 - \frac{(170)^2}{20} = 215$$

with $abn - 1 = 19$ d.f. For the treatment sum of squares we have

$$SS_T = \frac{(45)^2}{5} + \frac{(35)^2}{5} + \frac{(25)^2}{5} + \frac{(65)^2}{5} - \frac{(170)^2}{20} = 175$$

* Factorial experiments with unequal n's are discussed in Chapter 12.

with $ab - 1 = 3$ d.f.; by subtraction we obtain

$$SS_W = 215 - 175 = 40$$

with $ab(n - 1) = 16$ d.f.

If we combine the two levels of B for each level of A, we obtain the following sums: $\Sigma A_1 = 45 + 35 = 80$ and $\Sigma A_2 = 25 + 65 = 90$; each of these sums will be based on $n = 10$ observations. Then the sum of squares for A will be given by

$$SS_A = \frac{(80)^2}{10} + \frac{(90)^2}{10} - \frac{(170)^2}{20} = 5$$

with $a - 1 = 1$ d.f. In the same manner, we obtain $\Sigma B_1 = 45 + 25 = 70$ and $\Sigma B_2 = 35 + 65 = 100$; and each of these sums will be based on $n = 10$ observations. Then for the sum of squares for B we have

$$SS_B = \frac{(70)^2}{10} + \frac{(100)^2}{10} - \frac{(170)^2}{20} = 45$$

Table 9.2 shows a notation for the sample means for each of the A_iB_j treatment combinations, where the first subscript corresponds to the level of A and the second subscript corresponds to the level of B. For example, for treatment combination A_1B_1 the means is $\bar{Y}_{11}$. The mean for A_1, averaged over the two levels of B_j, is $\bar{Y}_{1.}$, and the mean for A_2, averaged over the two levels of B_j, is $\bar{Y}_{2.}$. Similarly, the mean

TABLE 9.2 Notation for a 2 × 2 factorial experiment

Sample				Population			
	B_1	B_2	Means		B_1	B_2	Means
A_1	$\bar{Y}_{11}$	$\bar{Y}_{12}$	$\bar{Y}_{1.}$	A_1	μ_{11}	μ_{12}	$\mu_{1.}$
A_2	$\bar{Y}_{21}$	$\bar{Y}_{22}$	$\bar{Y}_{2.}$	A_2	μ_{21}	μ_{22}	$\mu_{2.}$
Means	$\bar{Y}_{.1}$	$\bar{Y}_{.2}$	$\bar{Y}$	Means	$\mu_{.1}$	$\mu_{.2}$	μ

Values observed			
	B_1	B_2	Means
A_1	9.0	7.0	8.0
A_2	5.0	13.0	9.0
Means	7.0	10.0	8.5

for B_1 is $\bar{Y}_{.1}$, and the mean for B_2 is $\bar{Y}_{.2}$; the overall mean is $\bar{Y}$. The right-hand portion of Table 9.2 gives the corresponding notation for the population means. The bottom part gives the observed values of the means for the data in Table 9.1.

To calculate the interaction sum of squares we find

$$SS_{AB} = \Sigma n(\bar{Y}_{ij} - \bar{Y}_{i.} - \bar{Y}_{.j} + \bar{Y})^2 \tag{9.1}$$

where n is the number of observations for each treatment combination and the summation is over the $ab = 4$ values of $(\bar{Y}_{ij} - \bar{Y}_{i.} - \bar{Y}_{.j} + \bar{Y})^2$. Substituting with the values in Table 9.2 in (9.1), we obtain

$$\begin{aligned} 5(9.0 - 8.0 - 7.0 + 8.5)^2 &= 31.25 \\ 5(7.0 - 8.0 - 10.0 + 8.5)^2 &= 31.25 \\ 5(5.0 - 9.0 - 7.0 + 8.5)^2 &= 31.25 \\ 5(13.0 - 9.0 - 10.0 + 8.5)^2 &= 31.25 \end{aligned}$$

and

$$SS_{AB} = (4)(31.25) = 125$$

with 1 d.f. If a two-factor interaction sum of squares has 1 d.f., then we also have

$$SS_{AB} = \frac{n[\bar{Y}_{11} + \bar{Y}_{22}) - (\bar{Y}_{12} + \bar{Y}_{21})]^2}{4} \tag{9.2}$$

and substituting with the values given in Table 9.2, we also have

$$SS_{AB} = \frac{5[(9.0 + 13.0) - (7.0 + 5.0)]^2}{4} = 125$$

Because SS_A, SS_B, and SS_{AB} represent an orthogonal partitioning of SS_T, we have

$$SS_T = SS_A + SS_B + SS_{AB}$$

and, obviously, we could also have obtained SS_{AB} by subtraction, or

$$SS_{AB} = SS_T - SS_A - SS_B$$

9.3 Tests of Significance

Table 9.3 summarizes the analysis of variance. $F = MS_A/MS_W = 2.0$ is not significant, but $F = MS_B/MS_W = 18.0$ and $F = MS_{AB}/MS_W = 50.0$, each with 1 and 16 d.f., are both significant with $\alpha = .05$. The F test of MS_A is a test of significance of the difference between the means for A_1 and A_2 or for the difference between $\bar{Y}_{1.} = 8.0$ and $\bar{Y}_{2.} = 9.0$.

TABLE 9.3 Summary of the analysis of variance for the data in Table 9.1

Source of variation	Sum of squares	d.f.	Mean square	F
A	5.0	1	5.0	2.0
B	45.0	1	45.0	18.0
A × B	125.0	1	125.0	50.0
Within	40.0	16	2.5	50.0
Total	215.0	19		

The null hypothesis tested is that $\mu_{1.} = \mu_{2.}$. The F test of MS_B is a test of significance of the difference between the means for B_1 and B_2 or the difference between $\bar{Y}_{.1} = 7.0$ and $\bar{Y}_{.2} = 10.0$. The null hypothesis tested is that $\mu_{.1} = \mu_{.2}$.

The F test of MS_{AB} is a test of significance between two mean differences:

$$(\bar{Y}_{11} - \bar{Y}_{12}) - (\bar{Y}_{21} - \bar{Y}_{22}) = 10.0$$

or, equivalently,

$$(\bar{Y}_{11} - \bar{Y}_{21}) - (\bar{Y}_{12} - \bar{Y}_{22}) = 10.0$$

The null hypothesis tested is that

$$\mu_{11} - \mu_{12} = \mu_{21} - \mu_{22}$$

or, equivalently,

$$\mu_{11} - \mu_{21} = \mu_{12} - \mu_{22}$$

A significant $A \times B$ interaction indicates that the difference between the means for A_1 and A_2 depends on the level of B.* For example, with B_1, the difference between the means for A_1 and A_2 is $9.0 - 5.0 = 4.0$. But when we look at the difference between the means for A_1 and

* The test of significance of MS_A and MS_B may or may not be meaningful in the presence of a significant $A \times B$ interaction. With a significant $A \times B$ interaction, the difference between the means of A_1 and A_2 depends on or is conditional on the levels of B. The test of MS_A is a test of the difference between the means of A_1 and A_2 *averaged over the levels of* B. Thus, the test of MS_A is a test to determine whether there is a significant "overall A effect" in spite of or in addition to the $A \times B$ interaction.

In completely randomized designs involving human subjects, there is always the possibility of trait–treatment interactions. The test of MS_T in these experiments is a test in which the treatment means are averaged over countless individual difference variables, any one of which may interact significantly with the treatments. The fact that an individual difference variable has not been included in the design of the experiment and that it is ignored in the data analysis does not make the interaction disappear if, in fact, it exists. Thus, the test of MS_T in these experiments may be regarded as answering the question of whether there is a significant treatment effect in spite of or in addition to any interaction between the treatments and individual difference variables.

A_2 with B_2, we find that this difference is $7.0 - 13.0 = -6.0$, a reversal of the difference observed with B_1. Equivalently, the difference between the means of B_1 and B_2 depends on the level of A. For A_1, the difference between the means of B_1 and B_2 is $9.0 - 7.0 = 2.0$, but when we look at the difference between the means of B_1 and B_2 with A_2, we find that the difference is $5.0 - 13.0 = -8.0$.

9.4 Graphs for Two-Factor Interactions

In Figure 9.1 the means for A_1 and A_2 are plotted for each level of B. The fact that the difference between the two differences, $(9 - 5)$ and $(7 - 13)$, is equal to 10 rather than 0 results in the interaction sum of squares. An interaction of the kind shown in Figure 9.1 is called a *disordinal* interaction in that the two lines cross. When the two lines

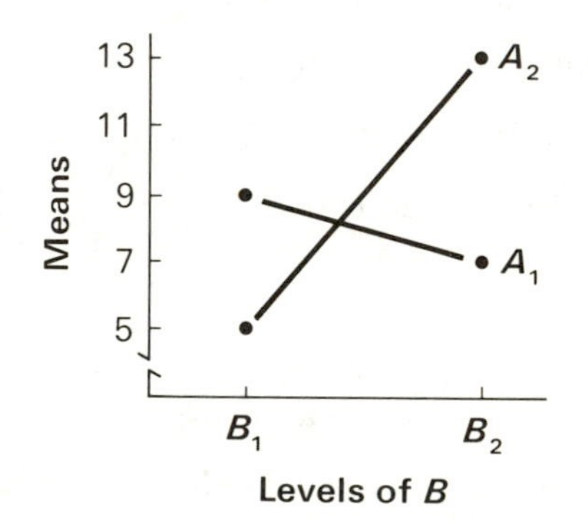

Figure 9.1 Graph of the $A \times B$ interaction in Table 9.3.

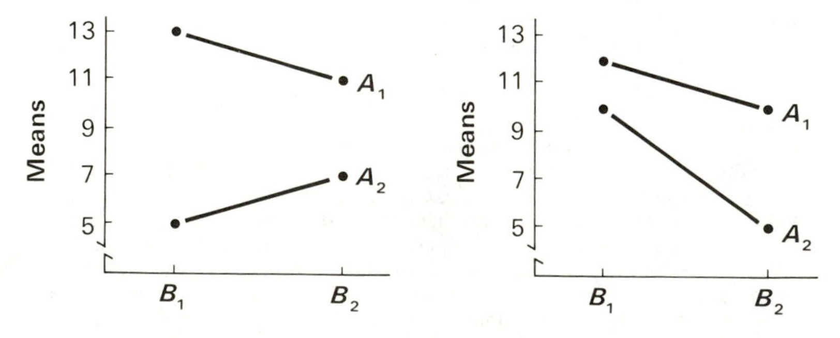

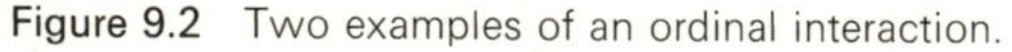

Figure 9.2 Two examples of an ordinal interaction.

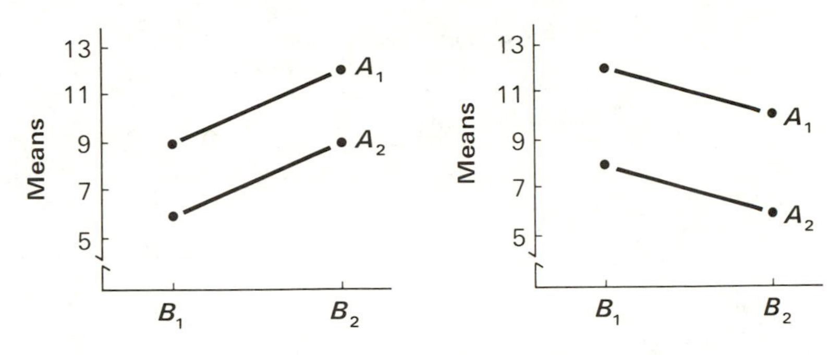

Figure 9.3 Two examples of a zero interaction.

cross, there is a reversal in the rank order of the means for A_1 and A_2 under the two levels of B.

Figure 9.2 illustrates *ordinal* interactions. The difference between the means of A_1 and A_2 is not the same for the two levels of B, but the two lines do not cross, and the rank order of the means is the same for both levels of B.* Figure 9.3 shows examples in which there is no interaction between A and B. The difference between the means for A_1 and A_2 remains the same for both levels of B.

9.5 Multiple Regression Analysis of the Data in Table 9.1

We turn now to a multiple regression analysis of the 2×2 factorial experiment. In Table 9.4 we have repeated the Y values for each of the $n = 5$ subjects in each of the A_iB_j treatments. Because we have four groups, we will need three X vectors, which we want to correspond to the A, B, and $A \times B$ effects of the analysis of variance. The first X vector, X_1, assigns 1 to each subject in A_1 and -1 to each subject in A_2. The second X vector, X_2, assigns 1 to each subject in B_1 and -1 to each subject in B_2. The third X vector, X_3, is simply the product of the corresponding entries in X_1 and X_2, or $X_3 = X_1X_2$. This vector carries the information about the interaction between A and B.

Observe that $\Sigma X_1 = \Sigma X_2 = \Sigma X_3 = 0$ so that the sum of squared deviations $\Sigma(X_i - \bar{X}_i)^2$ equals ΣX_i^2 for all X vectors. Observe also that the three X vectors are mutually orthogonal in that

$$\Sigma(X_i - \bar{X}_i)(X_j - \bar{X}_j) = \Sigma X_iX_j = 0$$

for all pairs of X vectors. Because the X vectors are mutually orthogonal, it will be true that

$$R^2_{Y.123} = r^2_{Y1} + r^2_{Y2} + r^2_{Y3}$$

* See, however, Exercise 9.5.

TABLE 9.4 Orthogonal coding for the data in Table 9.1

Treatments	X_1	X_2	X_3	Y
	1	1	1	7
	1	1	1	8
A_1B_1	1	1	1	9
	1	1	1	10
	1	1	1	11
	1	−1	−1	5
	1	−1	−1	6
A_1B_2	1	−1	−1	7
	1	−1	−1	8
	1	−1	−1	9
	−1	1	−1	3
	−1	1	−1	4
A_2B_1	−1	1	−1	5
	−1	1	−1	6
	−1	1	−1	7
	−1	−1	1	11
	−1	−1	1	12
A_2B_2	−1	−1	1	13
	−1	−1	1	14
	−1	−1	1	15

Because $\Sigma X_1 = 0$, we have $\Sigma x_1 y = \Sigma X_1 Y$, and it is easy to see, in Table 9.4, that

$$\Sigma x_1 y = \Sigma X_1 Y = (\Sigma Y_{11} + \Sigma Y_{12}) - (\Sigma Y_{21} + \Sigma Y_{22}) = -10$$

and

$$r_{Y1}^2 = \frac{(\Sigma x_1 y)^2}{(\Sigma x_1^2)(\Sigma y^2)} = \frac{(-10)^2}{(20)(215)} = .02326$$

Similarly,

$$\Sigma x_2 y = \Sigma X_2 Y = (\Sigma Y_{11} + \Sigma Y_{21}) - (\Sigma Y_{12} + \Sigma Y_{22}) = -30$$

and

$$r_{Y2}^2 = \frac{(\Sigma x_2 y)^2}{(\Sigma x_2^2)(\Sigma y^2)} = \frac{(-30)^2}{(20)(215)} = .20930$$

and

$$\Sigma x_3 y = \Sigma X_3 Y = (\Sigma Y_{11} + \Sigma Y_{22}) - (\Sigma Y_{12} + \Sigma Y_{21}) = 50$$

and

$$r_{Y3}^2 = \frac{(\Sigma x_3 y)^2}{(\Sigma x_3^2)(\Sigma y^2)} = \frac{(50)^2}{(20)(215)} = .58140$$

Then we have

$$R^2_{Y.123} = .02326 + .20930 + .58140 = .81396$$

Multiplying the above expression by SS_{tot}, we obtain

$$SS_{reg} = SS_{reg_1} + SS_{reg_2} + SS_{reg_3}$$

The left-hand side is the overall regression sum of squares, and the three terms on the right represent an orthogonal partitioning of SS_{reg}. We now note that

$$SS_{reg} = (215)(.81396) = (215)(.02326) + (215)(.20930) + (215)(.58140)$$

or

$$SS_{reg} = 175.0 = 5.0 + 45.0 + 125.0$$

and the overall regression sum of squares is equal to SS_T in the analysis of variance. The regression sums of squares for X_1, X_2, and X_3 correspond to the sums of squares for A, B, and the $A \times B$ interaction in the analysis of variance.

We also have, for the residual sum of squares,

$$SS_{res} = SS_{tot}(1 - R^2_{Y.123}) = (215)(1 - .81396) = 40.0$$

and SS_{res} is equal SS_W in the analysis of variance.

9.6 Tests of Significance with a Multiple Regression Analysis

For tests of significance with a multiple regression analysis we have

$$F = \frac{MS_{reg_i}}{MS_{res}} = F = \frac{r^2_{Yi}}{(1 - R^2_{Y.123})/[n - (k - 1) - 1]} \tag{9.3}$$

where $k - 1 = 3$ is the number of X vectors. To see the equivalence between the F tests defined by (9.3) and those in the analysis of variance, we calculate

$$F = \frac{.02326}{(1 - .81396)/16} = 2.0$$

which is equal to $F = MS_A/MS_W$ in the analysis of variance. Similarly,

$$F = \frac{.20930}{(1 - .81396)/16} = 18.0$$

which is equal to $F = MS_B/MS_W$ in the analysis of variance, and

$$F = \frac{.58140}{(1 - .81396)/16} = 50.0$$

which is equal to $F = MS_{AB}/MS_W$ in the analysis of variance.

It would also be easy to show that the F tests for b_1, b_2, and b_3 result in the same values of F that we obtained in testing the correlation coefficients for significance.

9.7 Analysis of Variance and Multiple Regression Analysis for a Factorial Experiment with Three Levels of A and Three Levels of B

In the analysis of variance of a factorial experiment with $a = 3$ levels of A and $b = 3$ levels of B with $n = 5$ subjects assigned at random to each of the $ab = 9$ treatment combinations, the total sum of squares, with 44 degrees of freedom, can be partitioned into the sum of squares within treatments with 36 d.f. and the treatment sum of squares with 8 d.f. The treatment sum of squares can then be partitioned into SS_A with 2 d.f., SS_B with 2 d.f., and SS_{AB} with 4 d.f.

Table 9.5 shows the X vectors that could be used in a multiple regression analysis of the same data. The first two X vectors, X_1 and X_2, provide an orthogonal partitioning of SS_A, and the second two X vectors, X_3 and X_4, correspond to an orthogonal partitioning of SS_B. Vectors X_5, X_6, X_7, and X_8 are product vectors for the interaction of A and B. For example, $X_5 = X_1X_3$, $X_6 = X_1X_4$, $X_7 = X_2X_3$, and $X_8 = X_2X_4$ are simply the products of the vectors for A and B. Observe that all eight X vectors are mutually orthogonal. Thus we have

$$R^2_{Y.12345678} = r^2_{Y1} + r^2_{Y2} + r^2_{Y3} + r^2_{Y4} + r^2_{Y5} + r^2_{Y6} + r^2_{Y7} + r^2_{Y8}$$

Then it will also be true that

$$R^2_{Y.12} = r^2_{Y1} + r^2_{Y2}$$

$$R^2_{Y.34} = r^2_{Y3} + r^2_{Y4}$$

and

$$R^2_{Y.5678} = r^2_{Y5} + r^2_{Y6} + r^2_{Y7} + r^2_{Y8}$$

TABLE 9.5 Orthogonal coding for a 3×3 factorial experiment

	X_1	X_2	X_3	X_4	X_5	X_6	X_7	X_8
A_1B_1	2	0	2	0	4	0	0	0
A_1B_2	2	0	−1	1	−2	2	0	0
A_1B_3	2	0	−1	−1	−2	−2	0	0
A_2B_1	−1	1	2	0	−2	0	2	0
A_2B_2	−1	1	−1	1	1	−1	−1	1
A_2B_3	−1	1	−1	−1	1	1	−1	−1
A_3B_1	−1	−1	2	0	−2	0	−2	0
A_3B_2	−1	−1	−1	1	1	−1	1	−1
A_3B_3	−1	−1	−1	−1	1	1	1	1

If these squared multiple correlation coefficients are multiplied by SS_{tot}, we will obtain

$$SS_{tot}R^2_{Y.12} = SS_A$$
$$SS_{tot}R^2_{Y.34} = SS_B$$

and

$$SS_{tot}R^2_{Y.5678} = SS_{AB}$$

Similarly, multiplying $1 - R^2_{Y.12345678}$ by SS_{tot}, we will find that

$$SS_{tot}(1 - R^2_{Y.12345678}) = SS_W$$

The total number of observations will be equal to nab, and the degrees of freedom for $1 - R^2_{Y.12345678}$ will be equal to

$$nab - (a - 1) - (b - 1) - (a - 1)(b - 1) - 1 = ab(n - 1)$$

The degrees of freedom for $R^2_{Y.12}$ will be equal to $a - 1$. Similarly, we will have $b - 1$ degrees of freedom for $R^2_{Y.34}$ and $(a - 1)(b - 1)$ degrees of freedom for $R^2_{Y.5678}$.

We let $R^2_{YY'} = R^2_{Y.12345678}$, and with $n = 5$, $a = 3$, and $b = 3$ we will have $(3)(3)(5 - 1) = 36$ d.f. for $1 - R^2_{YY'}$. Then, for the F test of $R_{Y.12}$ we will have

$$F = \frac{R^2_{Y.12}/2}{(1 - R^2_{YY'})/36} = \frac{MS_A}{MS_W}$$

Similarly, the F test for $R_{Y.34}$ is equivalent to the F test for MS_B in the analysis of variance, and the F test for $R_{Y.5678}$ is equivalent to the F test for MS_{AB} in the analysis of variance.

Exercises

9.1 Here is a simple example of a factorial experiment with two levels of A and two levels of B that can be used to repeat the analyses described in the chapter.

A_1B_1	A_1B_2	A_2B_1	A_2B_2
2	3	1	2
3	2	2	0
4	1	0	1

Complete the analysis of variance.

9.2 For the data in Exercise 9.1 construct three X vectors, one for A, one for B, and one for the $A \times B$ interaction, in the manner described in the chapter.

(a) Calculate r_{Y1}^2, r_{Y2}^2, and r_{Y3}^2. Are these values of r^2 equal to SS_A/SS_{tot}, SS_B/SS_{tot}, and SS_{AB}/SS_{tot}, respectively?

(b) What is the value of $R_{Y.123}^2$? Is it equal to SS_T/SS_{tot}?

(c) What is the value of $1 - R_{Y.123}^2$? Is it equal to SS_W/SS_{tot}?

(d) Test r_{Y1}, r_{Y2}, and r_{Y3} for significance. Are the F ratios obtained equal to MS_A/MS_W, MS_B/MS_W, and MS_{AB}/MS_W, respectively?

9.3 What is the difference between an interaction that is disordinal and an interaction that is ordinal?

9.4 For a factorial experiment with two levels of A and two levels of B, draw graphs in the manner described in the chapter to satisfy the conditions described:

(a) $SS_A = 0$, $SS_B = 0$, $SS_{AB} \neq 0$

(b) $SS_A \neq 0$, $SS_B = 0$, $SS_{AB} \neq 0$

(c) $SS_A \neq 0$, $SS_B = 0$, $SS_{AB} = 0$

9.5 We have a two-factor experiment with two levels of A and two levels of B with n subjects assigned at random to each of the four A_iB_j treatments. Suppose the means for the A_iB_j treatments are $\bar{Y}_{11} = 30$, $\bar{Y}_{12} = 20$, $\bar{Y}_{21} = 10$, and $\bar{Y}_{22} = 15$.

(a) Plot the means for the levels of A_1 and A_2 against the levels of B. Is the interaction ordinal or disordinal?

(b) Plot the means for the levels of B_1 and B_2 against the levels of A. Is the interaction ordinal or disordinal?

(c) For which of the two graphs shown in Figure 9.2 in the chapter is the interaction disordinal when the means for B_1 and B_2 are plotted against the levels of A?

9.6 Explain each of the following concepts:

balanced orthogonal design	interaction effect
disordinal interaction	levels of a factor
factor	ordinal interaction
factorial experiment	

10

A Repeated-Measure Design

10.1 Introduction

In all of the experiments described previously, different subjects were assigned at random to each of the treatments so that for each subject we had only one value of the dependent variable Y. In the experiments to be described now, each subject is tested under each of the various treatments so that for each subject we have not just one value of Y but a value of Y for each treatment. Experimental designs of this kind are called *repeated-measure* designs.

Suppose, for example, that an experimenter is interested in the blood sugar level (Y) when dogs are injected with different amounts of insulin

(X). If each animal is to receive each of the different dosages, then we obviously need to have some assurance that the effects of a previous dosage have worn off before administering the next dosage. In other words, if we are interested in studying the relationship between blood sugar level and amount of insulin injected with a repeated-measure design, then we want to be sure that there are no carry-over effects from one dosage to the next. One way to accomplish this goal would be to separate the administration of the injections by a sufficient period so that any effect of a previous dosage would have worn off.

In some experiments the carry-over effects themselves are of interest. For example, we might be interested in the number of errors (Y) made by subjects in learning a list of paired associates when the subjects are given a series of trials (X). In this experiment we might expect the number of errors to decrease from trial to trial as a result of practice or learning on previous trials; thus, the cumulative carry-over effects of the previous trials would be of experimental interest. In experiments on memory we might have subjects learn a list of paired associates and then test for their retention of the list over successive time periods.

10.2 Analysis of Variance for a Repeated-Measure Design with $k = 2$ Treatments

Table 10.1 gives the values of Y for $k = 2$ treatments administered to each of $n = 10$ subjects. The two treatments are such that we have no reason to believe that there would be carry-over effects from one treatment to the next. The order of administration of the two treatments

TABLE 10.1 Values of Y for $n = 10$ subjects who were tested under each of $k = 2$ treatments

Subjects	T_1	T_2	Σ
1	5	3	8
2	8	4	12
3	5	6	11
4	6	5	11
5	10	6	16
6	6	4	10
7	8	8	16
8	7	5	12
9	8	6	14
10	9	3	12
Σ	72	50	122

was independently randomized for each subject, with s
receiving T_1 followed by T_2 and other subjects receivin
by T_1. In Table 10.1 the Y values have been reordered so that the
columns correspond to the two treatments.

For the analysis of variance of the data in Table 10.1, we proceed in the usual way to find the total sum of squares, the treatment sum of squares, and the within treatment sum of squares. For the total sum of squares we have

$$SS_{tot} = (5)^2 + (8)^2 + \cdots + (3)^2 - \frac{(122)^2}{20} = 71.8$$

with $kn - 1 = 19$ d.f. The sum of squares for treatments will be given by

$$SS_T = \frac{(72)^2}{10} + \frac{(50)^2}{10} - \frac{(122)^2}{20} = 24.2$$

with $k - 1 = 1$ d.f., and we obtain the sum of squares within treatments by subtraction, or

$$SS_W = 71.8 - 24.2 = 47.6$$

with $k(n - 1) = 18$ d.f.

With a repeated-measure design, the sum of squares within treatments consists of two components: One of these is called the *subject sum of squares* (SS_S) and the other is called the *subject × treatment sum of squares* (SS_{ST}). The subject sum of squares can be obtained by first adding the values of Y for each subject. These sums are shown at the right of Table 10.1. The sum of squares for subjects is a measure of the variation of the means of the subjects about the overall mean, just as the treatment sum of squares is a measure of the variation of the treatment means about the overall mean. The calculation of the sum of squares for subjects is analogous to the calculation of the sum of squares for treatments. In Table 10.1 the means for each subject are 4.0, 6.0, . . . , 6.0, and the overall mean is 6.1. Then for the subject sum of squares we have

$$SS_S = 2[(4.0 - 6.1)^2 + (6.0 - 6.1)^2 + \cdots + (6.0 - 6.1)^2] = 28.8$$

or, equivalently,

$$SS_S = \frac{(8)^2}{2} + \frac{(12)^2}{2} + \frac{(11)^2}{2} + \cdots + \frac{(12)^2}{2} - \frac{(122)^2}{20} = 28.8$$

with $n - 1 = 9$ d.f.

TABLE 10.2 Summary of the analysis of variance for the data in Table 10.1

Source of variation	Sum of squares	d.f.	Mean square	F
Subjects	28.8	9	3.200	
Treatments	24.2	1	24.200	11.58
$S \times T$	18.8	9	2.089	
Total	71.8	19		

Subtracting the sum of squares for subjects from the sum of squares within treatments, we obtain the subject × treatment sum of squares, or

$$SS_{ST} = SS_W - SS_S$$

and, for our example, we have

$$SS_{ST} = 47.6 - 28.8 = 18.8$$

with $(k - 1)(n - 1) = 9$ d.f. Dividing SS_{ST} by its degrees of freedom, we have

$$MS_{ST} = \frac{18.8}{9} = 2.089$$

and MS_{ST} is the error mean square for testing MS_T for significance with a repeated-measure design.

Table 10.2 summarizes the analysis of variance. For the test of significance of MS_T, we have $F = MS_T/MS_{ST} = 11.58$, which is a significant value with $\alpha = .01$ for 1 and 9 d.f. We conclude that the means for Treatments 1 and 2 differ significantly.

10.3 Multiple Regression Analysis of the Experiment with $k = 2$ Repeated Measures

Table 10.3 repeats the Y values in Table 10.1 along with two X vectors. Because we have two treatments, we will require only one X vector to represent the treatments. Instead of using 1 and 0 coding, we have used 1 and -1 to represent Treatment 1 and Treatment 2, respectively. The second X vector, X_2, gives the sum for each subject over the two treatments. Note that each sum appears twice in this X vector. Observe also that X_1 and X_2 are orthogonal.

Because $\Sigma X_1 = 0$, we have $\Sigma x_1 y = \Sigma X_1 Y$, and

$$\Sigma x_1 y = \Sigma X_1 Y = \Sigma Y_1 - \Sigma Y_2 = 72 - 50 = 22$$

TABLE 10.3 The Y values in Table 10.1 with a coded vector for treatments (X_1) and a sum vector (X_2)

Subjects	X_1	X_2	Y
1	1	8	5
2	1	12	8
3	1	11	5
4	1	11	6
5	1	16	10
6	1	10	6
7	1	16	8
8	1	12	7
9	1	14	8
10	1	12	9
1	−1	8	3
2	−1	12	4
3	−1	11	6
4	−1	11	5
5	−1	16	6
6	−1	10	4
7	−1	16	8
8	−1	12	5
9	−1	14	6
10	−1	12	3

and we also have $\Sigma x_1^2 = \Sigma X_1^2$. Then for the squared correlation between Y and X_1 we have

$$r_{Y1}^2 = \frac{(22)^2}{(20)(71.8)} = .33705$$

and we observe that

$$r_{Y1}^2 = \frac{SS_T}{SS_{tot}} = \frac{24.2}{71.8} = .33705$$

To obtain the correlation coefficient between Y and X_2, we first find

$$\begin{aligned}\Sigma X_2 Y &= (8)(5 + 3) + 12(8 + 4) + \cdots + (12)(9 + 3) \\ &= (8)^2 + (12)^2 + \cdots + (12)^2 \\ &= 1546\end{aligned}$$

We also have

$$\Sigma X_2 = 2\Sigma Y = (2)(122) = 244$$

Then we have

$$\Sigma x_2 y = \Sigma X_2 Y - \frac{(\Sigma X_2)(\Sigma Y)}{20} = 1546 - \frac{(244)(122)}{20} = 57.6$$

Note that $\Sigma x_2 y$ is equal to k times the subject sum of squares or $(2)(28.8) = 57.6$. This will, in general, be true for any sum vector X_k; that is,

$$\Sigma x_k y = kSS_S$$

where k is the number of repeated measures and SS_S is the subject sum of squares in the analysis of variance.

We have found that $\Sigma X_2 Y = (8)^2 + (12)^2 + \cdots + (12)^2$, and it is apparent that $\Sigma X_2^2 = 2\Sigma X_2 Y$, because each of the squares, $(8)^2$, $(12)^2, \ldots,$ $(12)^2$, will appear twice in ΣX_2^2. Thus, we have $\Sigma X_2^2 = (2)(1546) =$ 3092, and

$$\Sigma x_2^2 = 3092 - \frac{(244)^2}{20} = 3092 - 2976.8 = 115.2$$

Note that Σx_2^2 in this example is equal to k^2 times the subject sum of squares or $(2)^2(28.8) = 115.2$. This will, in general, be true for any sum vector X_k; that is,

$$\Sigma x_k^2 = k^2 SS_S$$

where k is the number of repeated measures and SS_S is the subject sum of squares in the analysis of variance.

For the squared correlation between Y and X_2 we now have

$$r_{Y2}^2 = \frac{(57.6)^2}{(115.2)(71.8)} = .40111$$

and we observe that

$$r_{Y2}^2 = \frac{SS_S}{SS_{tot}} = \frac{28.8}{71.8} = .40111$$

Because X_1 and X_2 are orthogonal, $r_{12} = 0$, and, therefore,

$$R_{Y.12}^2 = r_{Y1}^2 + r_{Y2}^2 = .33705 + .40111 = .73816$$

If we divide the error sum of squares, SS_{ST}, by SS_{tot}, we obtain

$$\frac{SS_{ST}}{SS_{tot}} = \frac{18.8}{71.8} = .26184$$

which is equal to $1 - R_{Y.12}^2 = 1 - .73816 = .26184$.

In a repeated-measure design with conventional methods of coding, we will require $k - 1$ X vectors to identify the treatments and $n - 1$ X vectors to identify the subjects. In general, in a repeated-measure design we will have n subjects with k observations for each subject. The degrees of freedom associated with $1 - R^2_{YY'}$ will then be

$$kn - (k - 1) - (n - 1) - 1 = (k - 1)(n - 1)$$

We have used a simpler method of coding for the repeated-measure design, in which X_2, the sum vector, replaces the conventional $n - 1$ subject vectors.* Thus, the correct degrees of freedom associated with the sum vector† will be $n - 1$.

For the F test of r^2_{Y1} we have

$$F = \frac{r^2_{Y1}}{(1 - R^2_{Y.12})/(k - 1)(n - 1)}$$

or

$$F = \frac{.33705}{(1 - .73816)/9} = 11.58$$

which is equal to $F = MS_T/MS_{ST}$ in the analysis of variance.

10.4 Analysis of Variance of a Repeated-Measure Design with $k = 4$ Treatments

In a maze involving 10 choice points, $n = 5$ rats were tested on each of $k = 4$ trials (treatments). The number of errors made by each rat on each trial are shown in Table 10.4. The time intervals between the four trials are equally spaced. The plot of the mean number of errors against trials is shown in Figure 10.1. Prior to collecting the data, the experimenter planned on testing the linear, quadratic, and cubic components of the trial (treatment) sum of squares for significance.

For the analysis of variance we have the following sums of squares:

$$SS_{tot} = (8)^2 + (7)^2 + \cdots + (2)^2 - \frac{(100)^2}{20} = 82.0$$

$$SS_T = \frac{(35)^2}{5} + \frac{(26)^2}{5} + \frac{(21)^2}{5} + \frac{(18)^2}{5} - \frac{(100)^2}{20} = 33.2$$

$$SS_W = 82.0 - 33.2 = 48.8$$

$$SS_S = \frac{(28)^2}{4} + \frac{(23)^2}{4} + \frac{(21)^2}{4} + \frac{(16)^2}{4} + \frac{(12)^2}{4} - \frac{(100)^2}{20} = 38.5$$

* This method of coding subjects by a single vector was first described by Pedhazur (1977).

† Computer output will also assign a single degree of freedom to the sum vector and will incorrectly report the degrees of freedom for $1 - R^2_{YY'}$. Consequently, the values of t or F will be incorrect and will need to be recalculated.

TABLE 10.4 Number of errors made by each of $n = 5$ rats on each of $k = 4$ successive trials

	Trials					
Subjects	1	2	3	4	Σ	Means
1	8	8	7	5	28	7.00
2	7	7	5	4	23	5.75
3	9	4	4	4	21	5.25
4	6	4	3	3	16	4.00
5	5	3	2	2	12	3.00
Σ	35	26	21	18	100	
Means	7.0	5.2	4.2	3.6		5.00

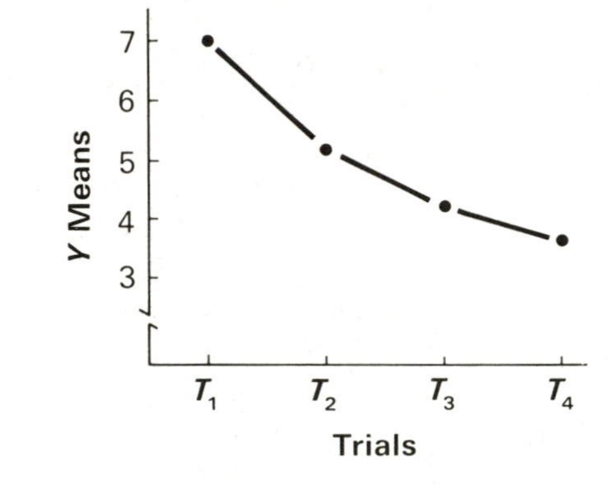

Figure 10.1 Mean number of errors made by $n = 5$ rats on each of $k = 4$ successive trials.

TABLE 10.5 Mean number of errors ($\bar{Y}_i$) on $k = 4$ successive trials and coefficients for orthogonal polynomials for the linear (X_1), quadratic (X_2), and cubic (X_3) components of SS_T

	Trials				
	1	2	3	4	
$\bar{Y}_i$	7.0	5.2	4.2	3.6	$\Sigma X_i \bar{Y}_i$
X_1	−3	−1	1	3	−11.2
X_2	1	−1	−1	1	1.2
X_3	−1	3	−3	1	−.4

and

$$SS_{ST} = 48.8 - 38.5 = 10.3$$

10.5 Linear, Quadratic, and Cubic Components of SS_T

Table 10.5 gives the means for each trial and also the coefficients for orthogonal polynomials for $k = 4$. The values of X_1, X_2, and X_3 were obtained from Table III in the appendix and correspond to the coefficients for the linear, quadratic, and cubic components of the trend of the trial means (or sums), respectively. The values of $\Sigma X_i \bar{Y}_i$ are given at the right of Table 10.5. If we square each of the values of $\Sigma X_i \bar{Y}_i$, multiply by $n = 5$, and divide by ΣX_i^2, we will obtain SS_L, SS_Q, and SS_C, respectively. Thus, we have

$$SS_L = \frac{n(\Sigma X_1 \bar{Y}_i)^2}{\Sigma X_1^2} = \frac{(5)(-11.2)^2}{20} = 31.36$$

$$SS_Q = \frac{n(\Sigma X_2 \bar{Y}_i)^2}{\Sigma X_2^2} = \frac{(5)(1.2)^2}{4} = 1.80$$

$$SS_C = \frac{n(\Sigma X_3 \bar{Y}_i)^2}{\Sigma X_3^2} = \frac{(5)(-.4)^2}{20} = .04$$

and we see that we have partitioned the trial sum of squares, SS_T, into three orthogonal components, each with 1 d.f., or

$$SS_T = SS_L + SS_Q + SS_C = 31.36 + 1.80 + .04 = 33.20$$

Almost all of the sum of squares for trials can be accounted for by the linear component, 31.36, and $F = MS_L/MS_{ST} = 31.36/.8583 = 36.54$ is significant with $\alpha = .01$ and with 1 and 12 d.f. None of the other components is significant.

10.6 Multiple Regression Analysis of the Data in Table 10.4

Table 10.6 shows the values of Y for each subject on each trial. The vectors X_1, X_2, and X_3 are coded with the coefficients for the linear, quadratic, and cubic components of SS_T, respectively. Again we have used a sum vector, X_4, for subject identification, and we must keep in mind that X_4 really represents $n - 1 = 4$ X vectors. Observe that all of the X vectors are mutually orthogonal and that $\Sigma X_1 = \Sigma X_2 = \Sigma X_3 = 0$.

To calculate r_{Y1}^2 we first calculate

$$\Sigma x_1 y = \Sigma X_1 Y = (-3)(35) + (-1)(26) + (1)(21) + (3)(18) = -56$$

TABLE 10.6 The Y values in Table 10.4 and X vectors coded with coefficients for orthogonal polynomials, X_1, X_2, and X_3, and a sum vector (X_4)

Subjects	X_1	X_2	X_3	X_4	Y
1	−3	1	−1	28	8
2	−3	1	−1	23	7
3	−3	1	−1	21	9
4	−3	1	−1	16	6
5	−3	1	−1	12	5
1	−1	−1	3	28	8
2	−1	−1	3	23	7
3	−1	−1	3	21	4
4	−1	−1	3	16	4
5	−1	−1	3	12	3
1	1	−1	−3	28	7
2	1	−1	−3	23	5
3	1	−1	−3	21	4
4	1	−1	−3	16	3
5	1	−1	−3	12	2
1	3	1	1	28	5
2	3	1	1	23	4
3	3	1	1	21	4
4	3	1	1	16	3
5	3	1	1	12	2

and we observe in Table 10.5 that $\Sigma X_1 \bar{Y}_i = -11.2$, and, consequently,

$$\Sigma X_1 Y = n\Sigma X_1 \bar{Y}_i = (5)(-11.2) = -56$$

Then we also have

$$\Sigma x_2 y = \Sigma X_2 Y = n\Sigma X_2 \bar{Y}_i = (5)(1.2) = 6.0$$
$$\Sigma x_3 y = \Sigma X_3 Y = n\Sigma X_3 \bar{Y}_i = (5)(-.4) = -2.0$$

We also have $\Sigma x_1^2 = \Sigma X_1^2 = (5)(20) = 100$, $\Sigma x_2^2 = \Sigma X_2^2 = (5)(4) = 20$, and $\Sigma x_3^2 = \Sigma X_3^2 = (5)(20) = 100$. Then with $SS_{tot} = 82$ we have

$$r_{Y1}^2 = \frac{(-56)^2}{(100)(82)} = .38244$$

$$r_{Y2}^2 = \frac{(6)^2}{(20)(82)} = .02195$$

$$r_{Y3}^2 = \frac{(-2)^2}{(100)(82)} = .00049$$

These values of r^2 are directly related to the values for SS_L, SS_Q, and SS_C that we calculated earlier. Thus,

$$r_{Y1}^2 = \frac{SS_L}{SS_{tot}} = \frac{31.36}{82.00} = .38244$$

$$r_{Y2}^2 = \frac{SS_Q}{SS_{tot}} = \frac{1.80}{82.00} = .02195$$

$$r_{Y3}^2 = \frac{SS_C}{SS_{tot}} = \frac{.04}{82.00} = .00049$$

We also need to calculate r_{Y4}^2, and to do this we first find

$$\begin{aligned} \Sigma X_4 Y &= (28)(8+8+7+5) + (23)(7+7+5+4) + \cdots \\ &\quad + (12)(5+3+2+2) \\ &= (28)^2 + (23)^2 + (21)^2 + (16)^2 + (12)^2 = 2154 \end{aligned}$$

We could calculate Σx_4^2, but, as we pointed out earlier in the chapter,

$$\Sigma x_4^2 = k^2 SS_S = (4)^2(38.5) = 616$$

We also observe that $\Sigma X_4 = 4\Sigma Y = (4)(100) = 400$. Then for $\Sigma x_4 y$ we have

$$\Sigma x_4 y = 2154 - \frac{(400)(100)}{20} = 154$$

which, as we know, must be equal to

$$\Sigma x_4 y = kSS_S = (4)(38.5) = 154$$

Then, for the value of r_{Y4}^2, we have

$$r_{Y4}^2 = \frac{(154)^2}{(616)(82)} = .46951$$

Because the X vectors are mutually orthogonal, we have

$$R_{Y.1234}^2 = r_{Y1}^2 + r_{Y2}^2 + r_{Y3}^2 + r_{Y4}^2$$

or

$$R_{Y.1234}^2 = .38244 + .02195 + .00049 + .46951 = .87439$$

We have $1 - R_{Y.1234}^2 = 1 - .87439 = .12561$, and we note that

$$1 - R_{Y.1234}^2 = \frac{SS_{ST}}{SS_{tot}} = \frac{10.3}{82.0} = .12561$$

Recall that the sum vector, X_4, in actuality corresponds to the $n - 1 = 4$ vectors that would be necessary to identify the subjects. Then the degrees of freedom for $1 - R^2_{Y.1234}$ will be equal to

$$kn - (k - 1) - (n - 1) - 1 = (k - 1)(n - 1)$$

or $(4 - 1)(5 - 1) = 12$, and

$$(1 - R^2_{Y.1234})/12 = .12561/12 = .01047$$

The F tests of r_{Y1}, r_{Y2}, and r_{Y3} will be equal, within rounding errors, to the F tests of MS_L, MS_Q, and MS_C, respectively, in the analysis of variance. For example, to test r_{Y1} for significance we have

$$F = \frac{.38244}{.01047} = 36.53$$

which is equal, within rounding errors, to $F = MS_L/MS_{ST}$.

10.7 Advantage of a Repeated-Measure Design

The advantage of a repeated-measure design over a completely randomized design is that MS_{ST} will usually be smaller than MS_W for the same number of observations. In fact, it can be shown that MS_{ST} is related to MS_W in the following way:

$$MS_{ST} = MS_W - \bar{c}_{ij}$$

where $\bar{c}_{ij}$ is the average of the $k(k - 1)$ covariances between the repeated measures. With only $k = 2$ repeated measures, $\bar{c}_{ij} = r_{12}s_1s_2$, where r_{12} is the correlation coefficient between the two repeated measures and s_1 and s_2 are the standard deviations of the measures for Treatment 1 and Treatment 2, respectively. For the data in Table 10.1, calculation would show that $r_{12}s_1s_2 = MS_W - MS_{ST}$. The reason that MS_{ST} will usually be smaller than MS_W is that, in general, the measures obtained from the same subjects will tend to be positively correlated so that $\bar{c}_{ij} = \overline{r_{ij}s_is_j}$ will be positive. Of course, the average value of the covariance must be sufficiently large to offset the fact that for the same number of observations, MS_{ST} will have fewer degrees of freedom than MS_W and will thus require a larger value of F for significance.

10.8 Assumptions in a Repeated-Measure Design

The advantage of a repeated-measure design is to some degree offset by the fact that an additional assumption is involved in the validity of the test of significance, $F = MS_T/MS_{ST}$, when we have $k > 2$ repeated measures. Recall that with independent values of Y for each treatment, the deviations of the Y values for a given treatment from the treatment mean, $e = Y_i - \bar{Y}_i$, were assumed to be independently and normally distributed with the same variance σ^2 in the population. The corresponding assumption in a multiple regression analysis involved the values of $e = Y - Y'$ for each fixed value of X.

The same assumption is involved in the analysis of variance of a repeated-measure design. A second and sufficient, but not necessary,* assumption for the validity of the test, $F = MS_T/MS_{ST}$, is that the correlation coefficients r_{ij} between the paired (Y_i, Y_j) measures for any two treatments, i and j, all have the same value ρ in the population.† If the Y measures for each treatment are independent of the Y measures for every other treatment, as is true when different subjects are randomly assigned to each of the treatments, then, in the population, the correlation between the Y measures for any two treatments will be equal to zero. With a repeated-measure design, the Y measures for the various treatments will, in general, not be independent but correlated. We thus need the additional assumption about equality of the correlation coefficients in the population. The assumption of equal variances and equal correlations is often referred to as the *compound symmetry assumption.*

Although a considerable amount is known about the robustness of the F test under violations of the assumption of equal variances, when the Y measures are independent, much less is known about the robustness of the F test in the case of a repeated-measure design under violations of the compound symmetry assumption. With equal n's for each treatment and with independent Y measures, the F test is quite robust against heterogeneity of variance. What little evidence there is available indicates that the F test in a repeated-measure design is robust against "moderate" violations of the compound symmetry assumption. It is difficult, of course, if not impossible, to define "moderate" precisely.

In the case of "strong" violations of the compound symmetry assumption, weaker, that is, less powerful, F tests can be made.‡ Other alternatives involve multivariate methods of analysis.

---

* Necessary *and* sufficient conditions are described by Huynh and Feldt (1970) and Rouanet and Lépine (1970). Both articles also describe tests to determine whether the necessary and sufficient conditions are violated.

† With $k = 2$ repeated measures, only one correlation coefficient is involved and there is no need to make any assumption about equality of the correlation coefficients.

‡ These F tests are described by Geisser and Greenhouse (1958) and Greenhouse and Geisser (1959).

Exercises

10.1 For the data in Table 10.3, calculate b_1. Is b_1 equal to $\bar{Y}_1 - \bar{Y}$? Explain why or why not.

10.2 For the data in Table 10.3, calculate b_2. Is the value of b_2 for the sum vector equal to $1/k$? Explain why or why not.

10.3 For the data in Table 10.3, calculate $a = \bar{Y} - b_1\bar{X}_1 - b_2\bar{X}_2$. Is a equal to zero? Explain why or why not.

10.4 For the data in Table 10.6, calculate the value of $a = \bar{Y} - b_1\bar{X}_1 - b_2\bar{X}_2 - b_3\bar{X}_3 - b_4\bar{X}_4$. Is the value of a equal to zero? Explain why $b_4\bar{X}_4$ is equal to $\bar{Y}$.

10.5 Here is a simple example of an experiment with three repeated measures:

	Treatments			
Subjects	T_1	T_2	T_3	Σ
1	5	6	6	17
2	4	4	8	16
3	3	5	4	12
4	2	7	7	16
5	1	3	5	9
Σ	15	25	30	70

Complete the analysis of variance.

10.6 A multiple regression analysis of the data in Exercise 10.5 will require two X vectors for the treatments. Let the values for X_1 be 2, -1, and -1. For X_2 let the values be 0, 1, and -1. To identify the subjects, use a sum vectors for X_3.

(a) Are the X vectors orthogonal?

(b) Calculate r^2_{Y1}, r^2_{Y2}, and r^2_{Y3}.

(c) Is $R^2_{Y.12}$ equal to SS_T/SS_{tot}?

(d) Is $1 - R^2_{Y.123}$ equal to SS_{ST}/SS_{tot}?

(e) Test $R_{Y.12}$ for significance. Is the value of F equal to $F = MS_T/MS_{ST}$ in the analysis of variance?

(f) Is the value of the Y intercept, a, equal to zero? Explain why or why not.

10.7 For the data in Exercise 10.5, do a multiple regression analysis with X_1 coded with the values 1, 0, and -1 and X_2 coded with the values 0, 1, and -1.

(a) Calculate r_{Y1}, r_{Y2}, and r_{Y3}.

(b) X_1 and X_2 are not orthogonal. Calculate r_{12} and then calculate $R^2_{Y.12}$, using a formula given in Section 4.4 for the two correlated X variables. Is $R^2_{Y.12}$ equal to the value obtained in Exercise 10.6?

(c) Calculate b_1 and b_2. Are they equal to $\bar{Y}_1 - \bar{Y}$ and $\bar{Y}_2 - \bar{Y}$, respectively?

(d) Calculate b_3 for the sum vector X_3. Is b_3 equal to $1/k$?

(e) Is the Y intercept, a, equal to zero? Explain why or why not.

10.8 For the Y values given in Exercise 10.5, calculate $r_{Y_1Y_2}$, $r_{Y_1Y_3}$, and $r_{Y_2Y_3}$. Then calculate s_1, s_2, and s_3, the standard deviations of the Y values for each treatment. The covariance between Y_i and Y_j will be given by $c_{ij} = r_{ij}s_is_j$, and there are three covariances involved. Calculate the three covariances and find the average value. Show that $MS_{ST} = MS_W - \overline{r_{ij}s_is_j}$.

10.9 We have k repeated measures for each of n subjects. We let ΣY_1, $\Sigma Y_2, \ldots, \Sigma Y_k$ be the sums for n subjects on the k repeated measures and $\Sigma Y = \Sigma Y_1 + \Sigma Y_2 + \cdots + \Sigma Y_k$. We let X_k be the sum vector. Prove that, in general:

(a) $\bar{X}_k = \Sigma Y/n$

(b) $\Sigma(X_k - \bar{X}_k)^2 = k^2 SS_S$

(c) $\Sigma x_k y = kSS_S$

(d) $b_k = 1/k$

10.10 Explain each of the following concepts:

carry-over effects

compound symmetry assumption

repeated-measure design

subject sum of squares

subject × treatment sum of squares

sum vector

11

Split-Plot Designs

11.1 Nature of a Split-Plot Design

In some experiments involving two factors, one factor may be of major interest and the other of minor interest. In this instance the experimenter may choose to use a design known as a *split-plot design*. We designate the factor of minor interest as A and the factor of major interest as B. In a split-plot design, subjects are randomly assigned to the levels of factor A, and then each subject in each level of A is tested under each of the levels of B. The split-plot design thus consists of a mixture of two designs: a completely randomized design with respect to the levels of A and a repeated-measure design with respect to the levels of B.

11.2 Analysis of Variance for a Split-Plot Design

The outcome of a simple example of a split-plot design is shown in Table 11.1. Five subjects were assigned at random to each of the two levels of A, and then each subject within each level of A was tested,

TABLE 11.1 Outcome for a split-plot design with $n = 5$ subjects randomly assigned to each level of A and with repeated measures on B

	Subjects	B_1	B_2	Σ
	1	1	3	4
	2	1	2	3
A_1	3	2	4	6
	4	6	5	11
	5	5	8	13
	Σ	15	22	37
	6	3	5	8
	7	5	4	9
A_2	8	6	7	13
	9	8	8	16
	10	8	11	19
	Σ	30	35	65

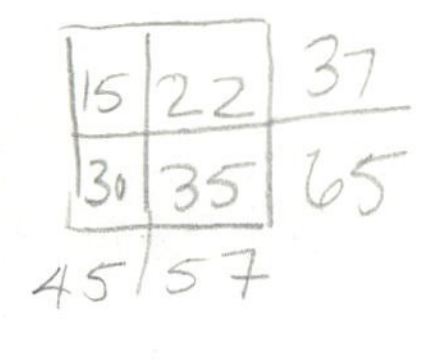

in random order, under each of the two levels of B. In Table 11.1 the values of Y have been rearranged so that the columns correspond to the two levels of B.

For the analysis of variance of the data in Table 11.1, we begin by calculating the total sum of squares, or

$$SS_{tot} = (1)^2 + (1)^2 + (2)^2 + \cdots + (11)^2 - \frac{(102)^2}{20} = 137.80$$

We then calculate the subject sum of squares, or

$$SS_S = \frac{(4)^2}{2} + \frac{(3)^2}{2} + \cdots + \frac{(19)^2}{2} - \frac{(102)^2}{20} = 120.80$$

The subject sum of squares is composed of two orthogonal components, the pooled sum of squares between subjects nested within each level of A, or $SS_{S(A)}$, and the sum of squares for A. For the sum for A_1 we have $15 + 22 = 37$, and for the sum for A_2 we have $30 + 35 = 65$. Then for the A sum of squares we have

$$SS_A = \frac{(37)^2}{10} + \frac{(65)^2}{10} - \frac{(102)^2}{20} = 39.20$$

and by subtraction we obtain

$$SS_{S(A)} = SS_S - SS_A \tag{11.1}$$

or

$$SS_{S(A)} = 120.80 - 39.20 = 81.60$$

The sum for B_1 is equal to $15 + 30 = 45$, and the sum for B_2 is equal to $22 + 35 = 57$; for the sum of squares for B we obtain

$$SS_B = \frac{(45)^2}{10} + \frac{(57)^2}{10} - \frac{(102)^2}{20} = 7.20$$

Modifying an equation we gave in Section 9.2 for calculating an interaction sum of squares with 1 d.f., we have

$$SS_{AB} = \frac{[(\Sigma A_1B_1 + \Sigma A_2B_2) - (\Sigma A_1B_2 + \Sigma A_2B_1)]^2}{4n} \tag{11.2}$$

and, for the data in Table 11.1, we obtain

$$SS_{AB} = \frac{[(15 + 35) - (22 + 30)]^2}{(4)(5)} = .20$$

We could calculate directly the pooled interaction sum of squares

$$SS_{S(A)B} = SS_{S(A_1)B} + SS_{S(A_2)B} \tag{11.3}$$

by considering first only those ten observations for A_1, or the top half of Table 11.1. For these observations, $SS_{S(A_1)B}$ corresponds to the subject × treatment sum of squares that we learned to calculate in the preceding chapter, that is, $SS_{S(A_1)B} = SS_{ST}$. Similarly, we could calculate $SS_{S(A_2)B} = SS_{ST}$ for the ten observations in A_2. We leave these calculations as an exercise and obtain $SS_{S(A)B}$ by subtraction. Thus, we have

$$SS_{S(A)B} = SS_{tot} - SS_S - SS_B - SS_{AB} \tag{11.4}$$

or, for the data in Table 11.1,

$$SS_{S(A)B} = 137.80 - 120.80 - 7.20 - .20 = 9.60$$

11.3 Summary of the Analysis of Variance

The results of our calculations are summarized in Table 11.2. There are two error mean squares in Table 11.2. The sum of squares for subjects, SS_S, has been partitioned into $SS_{S(A)}$ and SS_A. Recall that subjects were randomly assigned to the two levels of A, and the partitioning of SS_S corresponds to the partitioning of SS_{tot} in a completely randomized design; that is, $SS_{S(A)}$ corresponds to SS_W, and SS_A corresponds to SS_T. $MS_{S(A)}$ is the appropriate error mean square for testing the significance of MS_A. $MS_{S(A)B}$ corresponds to MS_{ST} for a repeated-measure

TABLE 11.2 Summary of the analysis of variance for the data in Table 11.1

Source of variation	Sum of squares	d.f.	Mean square	F
A	39.20	1	39.200	3.843
$S(A)$	81.60	8	10.200	
B	7.20	1	7.200	6.000
$A \times B$	.20	1	.200	
$S(A) \times B$	9.60	8	1.200	
Total	137.80	19		

design and is the appropriate error mean square for testing the significance of MS_B and MS_{AB}.* With 1 and 8 d.f., only $F = MS_B/MS_{S(A)B} = 6.00$ is significant with $\alpha = .05$. Note that $MS_{S(A)B}$ is smaller than $MS_{S(A)}$; this will generally be the case for a split-plot design. The F test for the significance of the factor of major interest, B, will ordinarily be a more sensitive test than the corresponding test of significance of the factor of minor interest, A.

11.4 Multiple Regression Analysis for a Split-Plot Design

In Table 11.3 the Y values for the split-plot design are repeated. Vector X_1 codes for the two levels of A, and X_2 codes for the two levels of B. The $A \times B$ interaction is carried by the product vector $X_3 = X_1X_2$; X_4 is the sum vector. Recall that the sum vector X_4 is simply a shortcut that avoids the need for nine X vectors to identify the $n = 10$ subjects. Note also that the sum vector X_4 will correspond to SS_S, and SS_S also includes SS_A, which we have represented by X_1.

Although the calculations of a multiple regression analysis of a split-plot design are accomplished easily by a computer, the computer output does not really help in understanding the relationships between the multiple regression analysis and the corresponding analysis of variance. That is the major reason that we have used a simple example in which

* For each level of the factor A to which subjects have been randomly assigned there will be a variance-covariance matrix of order $b \times b$, where b is the number of levels of the factor B for which we have repeated measures. A sufficient but not necessary condition for the validity of all F tests in a split-plot design is that the variance-covariance matrices for each level of A are equal in the population and satisfy the compound symmetry assumption. Necessary and sufficient conditions for the validity of all F tests are described by Huynh and Feldt (1970). They also describe tests for violations of the necessary and sufficient conditions. When the conditions for the validity of the usual F tests are not satisfied, weaker tests involving weaker assumptions are possible. These tests are described by Greenhouse and Geisser (1959).

TABLE 11.3 The Y values in Table 11.1 with orthogonal vectors, X_1, X_2, and X_3, for A, B, and $A \times B$, respectively, and a sum vector X_4

	Subjects	X_1	X_2	X_3	X_4	Y
	1	1	1	1	4	1
	2	1	1	1	3	1
A_1B_1	3	1	1	1	6	2
	4	1	1	1	11	6
	5	1	1	1	13	5
	1	1	−1	−1	4	3
	2	1	−1	−1	3	2
A_1B_2	3	1	−1	−1	6	4
	4	1	−1	−1	11	5
	5	1	−1	−1	13	8
	6	−1	1	−1	8	3
	7	−1	1	−1	9	5
A_2B_1	8	−1	1	−1	13	6
	9	−1	1	−1	16	8
	10	−1	1	−1	19	8
	6	−1	−1	1	8	5
	7	−1	−1	1	9	4
A_2B_2	8	−1	−1	1	13	7
	9	+1	−1	1	16	8
	10	−1	−1	1	19	11

the calculations can also be accomplished easily with a hand calculator. It is much easier to see the correspondence between the two methods of analysis when we actually show the multiple regression calculations.

Note first that the first three X vectors are mutually orthogonal and that we also have $\Sigma X_1 = \Sigma X_2 = \Sigma X_3 = 0$. Although X_4 is orthogonal with X_2 and X_3 so that $\Sigma X_2X_4 = \Sigma X_3X_4 = 0$, X_4 and X_1 are not orthogonal; that is, $\Sigma X_1X_4 \neq 0$. Vectors X_1 and X_4 are correlated. Then, for X_1, which codes for A, we have

$$\Sigma x_1 y = \Sigma X_1 Y = \Sigma A_1 - \Sigma A_2 = 37 - 65 = -28$$

which is just the difference between the sums for A_1 and A_2. With $\Sigma x_1^2 = \Sigma X_1^2 = 20$ and $\Sigma y^2 = SS_{tot} = 137.8$ we have

$$r_{Y1}^2 = \frac{(-28)^2}{(20)(137.8)} = .28447$$

and

$$r_{Y1}^2 = \frac{SS_A}{SS_{tot}} = \frac{39.2}{137.8} = .28447$$

Similarly, for X_2, which codes for the levels of B, we have

$$\Sigma x_2 y = \Sigma X_2 Y = \Sigma B_1 - \Sigma B_2 = 45 - 57 = -12$$

which is just the difference between the sums for B_1 and B_2. Then we have

$$r_{Y2}^2 = \frac{(-12)^2}{(20)(137.8)} = .05225$$

and

$$r_{Y2}^2 = \frac{SS_B}{SS_{tot}} = \frac{7.2}{137.8} = .05225$$

Similarly, for X_3, the product vector that codes for the $A \times B$ interaction, we have

$$\Sigma x_3 y = \Sigma X_3 Y = (\Sigma A_1 B_1 + \Sigma A_2 B_2) - (\Sigma A_1 B_2 + \Sigma A_2 B_1) = -2$$

and we note that this is simply the numerator of (11.2) for the $A \times B$ interaction sum of squares. Then we have

$$r_{Y3}^2 = \frac{(-2)^2}{(20)(137.8)} = .00145$$

and

$$r_{Y3}^2 = \frac{SS_{AB}}{SS_{tot}} = \frac{.20}{137.8} = .00145$$

Finally, we consider X_4, the sum vector. Observe that

$$\begin{aligned} \Sigma X_4 Y &= (4)(1 + 3) + (3)(1 + 2) + \cdots + (19)(8 + 11) \\ &= (4)^2 + (3)^2 + \cdots + (19)^2 \\ &= 1282 \end{aligned}$$

and that

$$\Sigma X_4 = 2\Sigma Y = (2)(102) = 204$$

Then we have

$$\Sigma x_4 y = 1282 - \frac{(204)(102)}{20} = 241.6$$

We pointed out in the preceding chapter that if X_k is a sum vector, then $\Sigma x_k y$ will be equal to kSS_S, where k is the number of repeated measures and SS_S is the subject sum of squares. In the example we are considering, we have $k = 2$ repeated measures and we found that SS_S

was equal to 120.8, so that we should also have

$$\Sigma x_4 y = (2)(120.8) = 241.6$$

We also know that if X_k is a sum vector, then $\Sigma x_k^2 = k^2 SS_S$. With $k = 2$ repeated measures and with $SS_S = 120.8$, we have $\Sigma x_k^2 = (2)^2(120.8) = 483.2$. As a check we also calculate $\Sigma x_k^2 = \Sigma X_k^2 - (\Sigma X_k)^2/n$. We have seen that

$$\Sigma X_4 Y = (4)^2 + (3)^2 + \cdots + (19)^2 = 1282$$

and it is easy to see that ΣX_4^2 is just two times $\Sigma X_4 Y$, or

$$\Sigma X_4^2 = 2\Sigma X_4 Y$$

or, for our example,

$$\Sigma X_4^2 = (2)(1282)$$

Then we have

$$\Sigma x_4^2 = (2)(1282) - \frac{(204)^2}{20} = 483.2$$

and Σx_4^2 is, in fact, equal to $k^2 SS_S$.

For the squared correlation between Y and X_4 we obtain

$$r_{Y4}^2 = \frac{(241.6)^2}{(483.2)(137.8)} = .87663$$

and

$$r_{Y4}^2 = \frac{SS_S}{SS_{tot}} = \frac{120.8}{137.8} = .87663$$

We pointed out earlier in the chapter that X_4, the sum vector, also carries the information regarding A that is coded by X_1. In fact,

$$r_{Y4}^2 = \frac{SS_S}{SS_{tot}} = \frac{SS_A}{SS_{tot}} + \frac{SS_{S(A)}}{SS_{tot}}$$

We did not include the four X vectors that would have been needed to differentiate between the five subjects in A_1 and the four X vectors that would have been necessary to differentiate between the five subjects in A_2, but instead only the single sum vector X_4 to carry all of the variation between subjects. In addition, X_4 carries the variation attributable to A. The correlation between Y and X_4 corresponds, therefore, to a multiple correlation between Y and a set of $4 + 4 + 1 = 9$ X vectors.

The squared semipartial correlation between Y and X_4 with X_1 partialed from X_4 will be given by

$$r_{Y(4.1)}^2 = r_{Y4}^2 - r_{Y1}^2 = .87663 - .28447 = .59216$$

and $r^2_{Y(4.1)}$ will have $9 - 1 = 8$ d.f. We now observe that

$$r^2_{Y(4.1)} = \frac{SS_{S(A)}}{SS_{tot}} = \frac{81.6}{137.8} = .59216$$

Because X_2, X_3, and X_4 are mutually orthogonal, we have

$$\begin{aligned} R^2_{Y.234} &= r^2_{Y2} + r^2_{Y3} + r^2_{Y4} \\ &= .05225 + .00145 + .87663 \\ &= .93033 \end{aligned}$$

and these squared correlation coefficients give the proportion of the total sum of squares accounted for by X_2, or SS_B; X_3, or SS_{AB}; and X_4, or $SS_A + SS_{S(A)} = SS_S$. Then

$$1 - R^2_{Y.234} = 1 - .93033 = .06967$$

and $1 - R^2_{Y.234} = .06967$ is the multiple regression counterpart of (11.4) in the analysis of variance. Thus we have

$$1 - R^2_{Y.234} = \frac{SS_{S(A)B}}{SS_{tot}} = \frac{9.6}{137.8} = .06967$$

Recall that X_4 has 9 d.f. and that X_2 and X_3 each have 1 d.f., so $1 - R^2_{Y.234}$ will have $n - k - 1 = 20 - 11 - 1 = 8$ d.f.

Summarizing the various calculations, we see that:

$$\begin{aligned} SS_A &= SS_{tot} r^2_{Y1} = (137.8)(.28447) = 39.2 \\ SS_{S(A)} &= SS_{tot} r^2_{Y(4.1)} = (137.8)(.59216) = 81.6 \\ SS_B &= SS_{tot} r^2_{Y2} = (137.8)(.05225) = 7.2 \\ SS_{AB} &= SS_{tot} r^2_{Y3} = (137.8)(.00145) = .2 \\ SS_{S(A)B} &= SS_{tot}(1 - R^2_{Y.234}) = (137.8)(1 - .93033) = 9.6 \end{aligned}$$

11.5 Tests of Significance with a Multiple Regression Analysis

For tests of significance with the multiple regression analysis we have

$$F = \frac{.28447}{.59216/8} = 3.843 = \frac{MS_A}{MS_{S(A)}}$$

$$F = \frac{.05225}{.06967/8} = 6.000 = \frac{MS_B}{MS_{S(A)B}}$$

and

$$F = \frac{.00145}{.06967/8} = .167 = \frac{MS_{AB}}{MS_{S(A)B}}$$

TABLE 11.4 A simple example of a split-plot design

	Subjects	B_1	B_2	Σ
	1	1	7	8
A_1	2	1	5	6
	3	2	6	8
	4	3	4	7
A_2	5	2	5	7
	6	4	3	7
	Σ	13	30	43

TABLE 11.5 Summary of the analysis of variance for the data in Table 11.4

Source of variation	Sum of squares	d.f.	Mean square	F
A	.083	1	.083	.249
$S(A)$	1.333	4	.333	
B	24.083	1	24.083	18.067
$A \times B$	10.083	1	10.083	7.564
$S(A) \times B$	5.333	4	1.333	
Total	40.915	11		

11.6 Multiple Regression Analysis with Subject Vectors

In order to show the greater efficiency of using a single sum vector in a multiple regression analysis for a repeated-measure design rather than individual subject X vectors, we consider a simple example. Table 11.4 gives the outcome of a split-plot design with three subjects assigned at random to each of two levels of A. Each subject in each level of A is tested under each level of B. The analysis of variance for the data in Table 11.4 follows the same pattern as for the example we have just discussed; we strongly recommend that you do the rather simple calculations involved. The calculations are summarized in Table 11.5*.

* In this example, $MS_{S(A)} = .333$ is smaller than $MS_{S(A)B} = 1.333$, which is contrary to the usual situation, in which $MS_{S(A)}$ is larger than $MS_{S(A)B}$. It can be shown that $MS_{S(A)} - MS_{S(A)B} = k\bar{c}_{ij}$, where k is the number of repeated measures and $\bar{c}_{ij}$ is the *average* of the covariances for the observations in each level of A. Typically, $\bar{c}_{ij}$ will be positive, and $MS_{S(A)}$ will be larger than $MS_{S(A)B}$. In this contrived example, $\bar{c}_{ij}$ is equal to $-.50$, and, consequently, $MS_{S(A)}$ is smaller than $MS_{S(A)B}$. For a further discussion, see Exercises 11.7 and 11.8 at the end of the chapter.

TABLE 11.6 Orthogonal coding for the data in Table 11.4 using subject vectors instead of a sum vector

Subjects	X_1	X_2	X_3	X_4	X_5	X_6	X_7	X_8	X_9	X_{10}	X_{11}	Y
$S_1(A_1)$	2	0	0	0	1	1	1	2	0	0	0	1
$S_2(A_1)$	−1	1	0	0	1	1	1	−1	1	0	0	1
$S_3(A_1)$	−1	−1	0	0	1	1	1	−1	−1	0	0	2
$S_4(A_2)$	0	0	2	0	−1	1	−1	0	0	2	0	3
$S_5(A_2)$	0	0	−1	1	−1	1	−1	0	0	−1	1	2
$S_6(A_2)$	0	0	−1	−1	−1	1	−1	0	0	−1	−1	4
$S_1(A_1)$	2	0	0	0	1	−1	−1	−2	0	0	0	7
$S_2(A_1)$	−1	1	0	0	1	−1	−1	1	−1	0	0	5
$S_3(A_1)$	−1	−1	0	0	1	−1	−1	1	1	0	0	6
$S_4(A_2)$	0	0	2	0	−1	−1	1	0	0	−2	0	4
$S_5(A_2)$	0	0	−1	1	−1	−1	1	0	0	1	−1	5
$S_6(A_2)$	0	0	−1	−1	−1	−1	1	0	0	1	1	3

$S(A_1)$	X_1 and X_2
$S(A_2)$	X_3 and X_4
A	X_5
B	X_6
$A \times B$	$X_7 = X_5X_6$
$S(A_1) \times B$	$X_8 = X_1X_6$ and $X_9 = X_2X_6$
$S(A_2) \times B$	$X_{10} = X_3X_6$ and $X_{11} = X_4X_6$

In Table 11.6 there is no sum vector. Instead we have used two X vectors, X_1 and X_2, to identify the three subjects in A_1 and two X vectors, X_3 and X_4, to identify the three subjects in A_2. X_5 is a vector representing the two levels of A, and X_6 is a vector representing the two levels of B. The interaction, $A \times B$, is carried by the product vector $X_7 = X_5X_6$. We also need product vectors to carry $S(A_1) \times B$; these are given by $X_8 = X_1X_6$ and $X_9 = X_2X_6$. Similarly, for $S(A_2) \times B$ we have the two product vectors $X_{10} = X_3X_6$ and $X_{11} = X_4X_6$. For each of the X vectors we have $\Sigma X = 0$, and all of the X vectors are mutually orthogonal; that is, we have $\Sigma X_iX_j = 0$ for all pairs of X vectors. Furthermore, we see that $\Sigma X_3Y = \Sigma X_4Y = \Sigma X_9Y = \Sigma X_{10}Y = 0$, so we know that $r_{Y3} = r_{Y4} = r_{Y9} = r_{Y10} = 0$.

Table 11.7 shows the values of r_{Yi}^2 between Y and each X vector and also the increase in $R_{YY'}^2$ as each X vector is added to the regression equation. Because X_1, X_2, X_3, and X_4 jointly represent the subjects within each level of A, we have

$$SS_{S(A)} = SS_{tot}R_{Y.1234}^2 = (40.915)(.03259) = 1.333$$

If we add in X_5 representing the A effect, we have as the increment

TABLE 11.7 Values of $r^2_{Y_i}$ and increase in R^2 as each X vector is added to the regression equation. The X vectors are those shown in Table 11.6

Vector	$r^2_{Y_i}$	Increase in R^2
X_1	.00815	.00815
X_2	.02444	.03259
X_3	.00000	.03259
X_4	.00000	.03259
X_5	.00203	.03462
X_6	.58860	.62322
X_7	.24643	.86965
X_8	.03259	.90224
X_9	.00000	.90224
X_{10}	.00000	.90224
X_{11}	.09776	1.00000

in $R^2_{YY'}$ due to A the difference between $R^2_{Y.12345}$ and $R^2_{Y.1234}$, or

$$SS_A = SS_{tot}(R^2_{Y.12345} - R^2_{Y.1234}) = (40.915)(.00203) = .083$$

Similarly,

$$SS_B = SS_{tot}(R^2_{Y.123456} - R^2_{Y.12345}) = (40.915)(.58860) = 24.083$$

and

$$SS_{AB} = SS_{tot}(R^2_{Y.1234567} - R^2_{Y.123456}) = (40.915)(.24643) = 10.083$$

Finally, because X_8, X_9, X_{10}, and X_{11} jointly carry the interaction $S(A) \times B$, we have*

$$SS_{S(A)B} = SS_{tot}(R^2_{Y.123\ldots11} - R^2_{Y.123\ldots7}) = (40.915)(.13035) = 5.333$$

If we had used a sum vector in this analysis instead of vectors identifying the subjects in each level of A, we would have used only four X vectors. We leave this analysis for one of the exercises.

---

* It should come as no surprise that $R^2_{Y.123\ldots11}$, in this example, is equal to one, because we have only $n = 12$ values of Y. Recall that $R^2_{YY'}$ has $n - k - 1$ degrees of freedom, where n is the number of values of Y, and k is the number of X vectors. Consequently, any set of $k = n - 1$ orthogonal X vectors will always result in $R^2_{YY'} = 1.0$. In fact, the $n - 1$ X vectors need not be orthogonal in order for $R^2_{YY'}$ to be equal to one. It is only necessary that each X vector have a variance greater than zero and that no X vector have a multiple correlation of one with the remaining X vectors. If these conditions are satisfied, then $R^2_{YY'}$ must be equal to one.

In this example, our only interest in $R^2_{Y.123\ldots11}$ was to obtain $R^2_{Y.123\ldots11} - R^2_{Y.123\ldots7}$, and, because we knew that $R^2_{Y.123\ldots11}$ had to be equal to one, we could have omitted vector X_8 through X_{11}. We retained these X vectors because we wanted the reader to see that $R^2_{Y.123\ldots11}$ would, in fact, be equal to one.

Exercises

11.1 In what respect is a split-plot design a mixture of a completely randomized design and a repeated-measure design?

11.2 If we have two factors of interest, but one is of major interest and the other is of minor interest, for which factor should we obtain repeated measures? Explain why.

11.3 Suppose that we have a split-plot design with five subjects assigned completely at random to the three levels of factor A. Each subject is tested under each of the three levels of another factor B; that is, we have $k = 3$ repeated measures for each subject. Make a table showing the various sources of variation and the degrees of freedom associated with each source.

11.4 For the example in Exercise 11.3, suppose that we do a multiple regression analysis using a sum vector.
(a) What is the total number of X vectors needed, and what will they correspond to?
(b) How many degrees of freedom will be associated with the sum vector?
(c) How could we obtain $SS_{S(A)}$ from the multiple regression analysis?

11.5 For the example in Exercise 11.3, suppose that we do a multiple regression analysis without a sum vector. Instead we use separate X vectors to identify the subjects within each level of A.
(a) How many subject vectors will we need?
(b) Altogether, how many X vectors will be involved in this analysis?
(c) How could we obtain $SS_{S(A)}$ from the multiple regression analysis?

11.6 In the chapter we did a multiple regression analysis of the data in Table 11.4, using subject vectors.
(a) Repeat the analysis using a sum vector instead of subject vectors.
(b) Compare your results with those reported in the text.

11.7 For the data in Table 11.1, calculate the *average* of the covariances for the observations in A_1 and the observations in A_2. You should find that $MS_{S(A)} - MS_{S(A)B} = k\bar{c}_{ij}$, where $k = 2$ is the number of repeated measures and $\bar{c}_{ij}$ is the average of the covariances.

11.8 It can be shown that $MS_{S(A)B} = MS_W - \bar{c}_{ij}$, where MS_W is the average of the variances between subjects in each level of A_iB_j and $\bar{c}_{ij}$ is the average of the covariances for each level of A. It can also be shown that $MS_{S(A)} = MS_W + (k - 1)\bar{c}_{ij}$, where k is the number of repeated measures.
(a) Calculate MS_W for the data in Table 11.1, and show that $MS_{S(A)B} = MS_W - \bar{c}_{ij}$ and $MS_{S(A)} = MS_W + (k - 1)\bar{c}_{ij}$.
(b) Subtract $MS_{S(A)B}$ from $MS_{S(A)}$, and show that $MS_{S(A)} - MS_{S(A)B} = k\bar{c}_{ij}$.

11.9 Suppose we have $n = 3$ values of a Y variable and two X variables as shown below:

X_1	X_2	Y
3	0	1
0	3	2
0	0	3

Calculate

$$R^2_{Y.12} = \frac{r^2_{Y1} + r^2_{Y2} - 2r_{Y1}r_{Y2}r_{12}}{1 - r^2_{12}}$$

11.10 Explain why the correlation of Y with a sum vector is, in essence, a multiple correlation.

11.11 Explain each of the following concepts:

split-plot design sum vector

subject vector

12

Nonorthogonal Designs: Two-Factor Experiments

12.1 Orthogonal Designs

Suppose that we have a factorial experiment with $a = 3$ levels of A and $b = 4$ levels of B and with n subjects assigned at random to each of the A_iB_j treatment combinations. The cells of Table 12.1 give the means for each of the A_iB_j treatment combinations. The row means are the means for the levels of A, or the means for A_1, A_2, and A_3. The column means are the means for the levels of B, or the means for B_1, B_2, B_3, and B_4.

If we let $\bar{Y}_{ij}$ represent a cell mean, $\bar{Y}$ represent the overall mean, $\bar{Y}_i$ represent the mean for the ith level of A, and $\bar{Y}_j$ represent the mean for the jth level of B, we have the following identity:

$$\bar{Y}_{ij} = \bar{Y} + (\bar{Y}_i - \bar{Y}) + (\bar{Y}_j - \bar{Y}) + (\bar{Y}_{ij} - \bar{Y}_i - \bar{Y}_j + \bar{Y}) \qquad (12.1)$$

where the terms in the parentheses on the right correspond to the A effect, the B effect, and the $A \times B$ interaction.

If we let Y_{ijk} correspond to an observation obtained from treatment combination A_iB_j, then it follows from (12.1) that

$$Y_{ijk} - \bar{Y} = (Y_{ijk} - \bar{Y}_{ij}) + (\bar{Y}_i - \bar{Y}) + (\bar{Y}_j - \bar{Y}) + (\bar{Y}_{ij} - \bar{Y}_i - \bar{Y}_j + \bar{Y}) \tag{12.2}$$

If we have an equal number of n observations for each treatment combination, then it can be shown that when (12.2) is squared and summed over all observations, all cross-product terms on the right disappear and only the sums of the squared terms remain. This is because with equal n's the design is orthogonal. Furthermore, the sums of the squared terms correspond exactly to

$$SS_{tot} = SS_W + SS_A + SS_B + SS_{AB}$$

Recall that SS_W in the analysis of variance corresponds to SS_{res} in the multiple regression analysis of the same data and that SS_T corresponds to SS_{reg}. With an orthogonal design it is possible to partition $SS_T = SS_{reg}$ into the orthogonal components SS_A, SS_B, and SS_{AB}.

12.2 Nonorthogonal Designs

Now suppose that for the experiment described in Table 12.1 we have missing observations for some of the treatment combinations. We assume that the missing observations are simply random losses and are in no way related to the nature of the treatments. More specifically, let us assume that the values of $\bar{Y}_{ij}$ are the same as those given in Table 12.1 but that we have unequal n's for the treatment combinations, as shown in Table 12.2. With unequal n's it is still possible in the analysis of variance to partition the total sum of squares into the sum

TABLE 12.1 Means for treatment combinations A_iB_j and for the levels of A_i averaged over the levels of B_j and for the levels of B_j averaged over the levels of A_i

	B_1	B_2	B_3	B_4	Means
A_1	6.00	7.00	4.50	3.50	5.250
A_2	4.50	4.00	8.00	7.50	6.000
A_3	9.00	4.00	6.50	9.00	7.125
Means	6.500	5.000	6.333	6.667	6.125

TABLE 12.2 Values of Y for a 3 × 4 factorial experiment with unequal n's. The means for each treatment combination are the same as those given in Table 12.1

Treatment combination	Values of Y	n	$\bar{Y}_{ij}$
A_1B_1	6, 7, 5	3	6.00
A_1B_2	7, 6, 8	3	7.00
A_1B_3	4, 5	2	4.50
A_1B_4	5, 3, 2, 4	4	3.50
A_2B_1	4, 5	2	4.50
A_2B_2	4, 3, 5	3	4.00
A_2B_3	9, 8, 7	3	8.00
A_2B_4	7, 8	2	7.50
A_3B_1	10, 8, 9	3	9.00
A_3B_2	5, 3	2	4.00
A_3B_3	7, 6	2	6.50
A_3B_4	10, 8, 9	3	9.00

of squares within treatments and the treatment sum of squares so that

$$SS_{tot} = SS_W + SS_T$$

Similarly, with a multiple regression analysis of the same data we have

$$SS_{tot} = SS_{res} + SS_{reg}$$

and it will be true that $SS_W = SS_{res}$ and $SS_T = SS_{reg}$. But with unequal n's there is no *unique* way in which to partition SS_T into SS_A, SS_B, and SS_{AB}. Similarly, with a multiple regression analysis of the same data there is no *unique* way in which to partition the overall regression sum of squares into a regression sum of squares for A, a regression sum of squares for B, and a regression sum of squares for the $A \times B$ interaction.

This does not mean that the situation is hopeless, but rather that a number of alternative methods for analyzing the data are possible and that these alternatives depend on various assumptions.* The method of analysis that we will describe is one that we believe is appropriate for the analysis of experimental data.†

We assume that the experimenter initially planned on having an equal number of observations for each treatment and that the missing observations are randomly missing and in no way related to the nature

* These assumptions are discussed briefly in Section 12.8 and more fully by Carlson and Timm (1974) and Lewis and Keren (1977). See also Overall and Spiegel (1969), Rawlings (1973), Joe (1971), Overall and Spiegel (1973), Applebaum and Cramer (1974), and Herr and Gaebelein (1978).

† We will explain why we believe the method to be appropriate in Section 12.8.

of the treatments. We also assume that if there were no missing observations, the experimenter would have partitioned the treatment sum of squares into the sum of squares for A, the sum of squares for B, and the sum of squares for the $A \times B$ interaction, and that the tests of significance of MS_A, MS_B, and MS_{AB} would all be made using MS_W as the common error mean square. Furthermore, if any of the mean squares failed to be significant, the experimenter would not assume that the null hypothesis had been proved and proceed to pool the sum of squares for nonsignificant mean squares with the error or within treatment sum of squares. This, we believe, is the procedure most experimenters would follow in analyzing the data of a factorial experiment with equal n's and with a fixed-effects model.

12.3 A Population Model for a Two-Factor Experiment

The population model corresponding to (12.1) is

$$\mu_{ij} = \mu + (\mu_i - \mu) + (\mu_j - \mu) + (\mu_{ij} - \mu_i - \mu_j + \mu) \qquad (12.3)$$

and does not depend on equal n's for the various treatment combinations. If we let $\bar{A}_i$ represent the mean for the ith level of A, $\bar{B}_j$ the mean for the jth level of B, and $\overline{A_iB_j}$ the mean for the treatment combination corresponding to the ith level of A and the jth level of B, then we can rewrite (12.1) as

$$\overline{A_iB_j} = \bar{Y} + (\bar{A}_i - \bar{Y}) + (\bar{B}_j - \bar{Y}) + (\overline{A_iB_j} - \bar{A}_i - \bar{B}_j + \bar{Y}) \qquad (12.4)$$

and the population model for (12.4) is often expressed as

$$\mu_{ij} = \mu + \alpha_i + \beta_j + \alpha_i\beta_j \qquad (12.5)$$

where $\alpha_i = \mu_i - \mu$, $\beta_j = \mu_j - \mu$, and $\alpha_i\beta_j = \mu_{ij} - \mu_i - \mu_j + \mu$, with the side conditions that*

$$\Sigma\alpha_i = \Sigma\beta_j = \sum_i \alpha_i\beta_j = \sum_j \alpha_i\beta_j = 0 \qquad (12.6)$$

12.4 Effect Coding

With a multiple regression analysis it is possible to define X vectors for the $a - 1$ levels of A and the $b - 1$ levels of B in such a way that the side conditions of (12.6) are satisfied, even in the case of unequal n's. The coding scheme that we will use is sometimes referred to as

---

* This, we believe, is the relevant population model regardless of whether the design is orthogonal or nonorthogonal. It is what we have called Model 8 in Section 12.8.

effect coding. With this method of coding we assign 1 to all subjects in the first level of A, or A_1, and -1 to all subjects in the last level of A, or A_3, in X_1. All other subjects are assigned 0 in X_1. This vector is shown in Table 12.3. For X_2 we assign 1 to all subjects in the second level of A, or A_2, and -1 to all subjects in the last level of A, or A_3. All other subjects are assigned 0 in X_2. This X vector is also shown in Table 12.3. Note that these two X vectors define completely the $a = 3$ levels of A.

Similarly, for X_3 we assign 1 to all subjects in B_1 and -1 to all subjects in B_4. All other subjects are assigned 0 in X_3. For X_4 we assign 1 to all subjects in B_2 and -1 to all subjects in B_4. All other subjects are assigned 0 in X_4. For X_5 we assign 1 to all subjects in B_3 and -1 to all subjects in B_4. All other subjects are assigned 0. These X vectors are also shown in Table 12.3, and they define completely the $b = 4$ levels of B. The other X vectors, X_6 through X_{11}, shown in Table 12.3, are simply product vectors as indicated. Table 12.4 shows the values of the X vectors associated with each of the Y values in the experiment.

12.5 Estimates of the Model Parameters

Table 12.5 gives the values of the regression coefficient obtained with a sequential analysis in which the X vectors were entered into the regression equation in the order shown in Table 12.4. This sequential analysis will have no influence on the values of the regression coefficients when all X vectors are in the regression equation. The same values for the regression coefficients would be obtained if all of the X vectors were entered in some other sequence.

For the first two levels of A, A_1 and A_2, represented by X_1 and X_2, respectively, we have $b_1 = -.875$ and $b_2 = -.125$. The value of b_1 is simply the deviation of the mean $\bar{A}_1 = 5.250$ from the *unweighted* mean of the means, $\bar{Y} = 6.125$, as given in Table 12.1. Thus, $b_1 = \bar{A}_1 - \bar{Y} = 5.250 - 6.125 = -.875$ is an estimate of α_1. Similarly, the mean for A_2 is $\bar{A}_2 = 6.000$, and

$$b_2 = \bar{A}_2 - \bar{Y} = 6.000 - 6.125 = -.125$$

is an estimate of α_2. The missing estimate of α_3 can be obtained from the side condition that $\Sigma\alpha_i = 0$ and is, therefore, equal to *minus* the sum of b_1 and b_2. Because $b_1 + b_2 = -1.000$, the estimate of α_3 is 1.000. We note that the mean for A_3 is $\bar{A}_3 = 7.125$, and we also have

$$\bar{A}_3 = \bar{Y} = 7.125 - 6.125 = 1.000$$

which is, of course, an estimate of α_3.

TABLE 12.3 Effect coding for a 3 × 4 factorial experiment

	X_1	X_2	X_3	X_4	X_5	X_6	X_7	X_8	X_9	X_{10}	X_{11}
	A_1	A_2	B_1	B_2	B_3	A_1B_1	A_1B_2	A_1B_3	A_2B_1	A_2B_2	A_2B_3
A_1B_1	1	0	1	0	0	1	0	0	0	0	0
A_1B_2	1	0	0	1	0	0	1	0	0	0	0
A_1B_3	1	0	0	0	1	0	0	1	0	0	0
A_1B_4	1	0	−1	−1	−1	−1	−1	−1	0	0	0
A_2B_1	0	1	1	0	0	0	0	0	1	0	0
A_2B_2	0	1	0	1	0	0	0	0	0	1	0
A_2B_3	0	1	0	0	1	0	0	0	0	0	1
A_2B_4	0	1	−1	−1	−1	0	0	0	−1	−1	−1
A_3B_1	−1	−1	1	0	0	−1	0	0	−1	0	0
A_3B_2	−1	−1	0	1	0	0	−1	0	0	−1	0
A_3B_3	−1	−1	0	0	1	0	0	−1	0	0	−1
A_3B_4	−1	−1	−1	−1	−1	1	1	1	1	1	1

TABLE 12.4 Effect coding for the data in Table 12.2

X_1	X_2	X_3	X_4	X_5	X_6	X_7	X_8	X_9	X_{10}	X_{11}	Y
1	0	1	0	0	1	0	0	0	0	0	6
1	0	1	0	0	1	0	0	0	0	0	7
1	0	1	0	0	1	0	0	0	0	0	5
1	0	0	1	0	0	1	0	0	0	0	7
1	0	0	1	0	0	1	0	0	0	0	6
1	0	0	1	0	0	1	0	0	0	0	8
1	0	0	0	1	0	0	1	0	0	0	4
1	0	0	0	1	0	0	1	0	0	0	5
1	0	−1	−1	−1	−1	−1	−1	0	0	0	5
1	0	−1	−1	−1	−1	−1	−1	0	0	0	3
1	0	−1	−1	−1	−1	−1	−1	0	0	0	2
1	0	−1	−1	−1	−1	−1	−1	0	0	0	4
0	1	1	0	0	0	0	0	1	0	0	4
0	1	1	0	0	0	0	0	1	0	0	5
0	1	0	1	0	0	0	0	0	1	0	4
0	1	0	1	0	0	0	0	0	1	0	3
0	1	0	1	0	0	0	0	0	1	0	5
0	1	0	0	1	0	0	0	0	0	1	9
0	1	0	0	1	0	0	0	0	0	1	8
0	1	0	0	1	0	0	0	0	0	1	7
0	1	−1	−1	−1	0	0	0	−1	−1	−1	7
0	1	−1	−1	−1	0	0	0	−1	−1	−1	8
−1	−1	1	0	0	−1	0	0	−1	0	0	10
−1	−1	1	0	0	−1	0	0	−1	0	0	8
−1	−1	1	0	0	−1	0	0	−1	0	0	9
−1	−1	0	1	0	0	−1	0	0	−1	0	5
−1	−1	0	1	0	0	−1	0	0	−1	0	3
−1	−1	0	0	1	0	0	−1	0	0	−1	7
−1	−1	0	0	1	0	0	−1	0	0	−1	6
−1	−1	−1	−1	−1	1	1	1	1	1	1	10
−1	−1	−1	−1	−1	1	1	1	1	1	1	8
−1	−1	−1	−1	−1	1	1	1	1	1	1	9

TABLE 12.5 Values of the obtained regression coefficients and missing values based on the side conditions $\Sigma\alpha_i = \Sigma\beta_j = \Sigma_i\alpha_i\beta_j = \Sigma_i\alpha_i\beta_j = 0$

Vector	Effect	Regression coefficients	Missing values
X_1	α_1	−.875	
X_2	α_2	−.125	
	α_3		1.000
X_3	β_1	.375	
X_4	β_2	−1.125	
X_5	β_3	.208	
	β_4		.542
X_6	$\alpha_1\beta_1$	.375	
X_7	$\alpha_1\beta_2$	2.875	
X_8	$\alpha_1\beta_3$	−.958	
	$\alpha_1\beta_4$		−2.292
X_9	$\alpha_2\beta_1$	−1.875	
X_{10}	$\alpha_2\beta_2$	−.875	
X_{11}	$\alpha_2\beta_3$	1.792	
	$\alpha_2\beta_4$		.958
	$\alpha_3\beta_1$		1.500
	$\alpha_3\beta_2$		−2.000
	$\alpha_3\beta_3$		−.833
	$\alpha_3\beta_4$		1.333

It can easily be shown that the next three regression coefficients, .375, −1.125, and .208, based on the X vectors corresponding to the first three levels of B, or B_1, B_2, and B_3, respectively, are equal to the deviations of the means, $\bar{B}_1 = 6.500$, $\bar{B}_2 = 5.000$, and $\bar{B}_3 = 6.333$, from the mean $\bar{Y} = 6.125$. Again, because of the side condition $\Sigma\beta_j = 0$, the missing estimate of β_4 is simply minus the sum of the obtained estimates, or .542, and is equal to

$$\bar{B}_4 - \bar{Y} = 6.667 - 6.125 = .542$$

which is an estimate of β_4.

The missing values of regression coefficients for the interactions $\alpha_i\beta_j$ can be obtained in a similar manner because of the side conditions $\Sigma_i\alpha_i\beta_j = \Sigma_j\alpha_i\beta_j = 0$; that is, the values of $\alpha_i\beta_j$ must sum to zero over both rows and columns. The missing estimates are also given in Table 12.5. We note that the regression coefficient for the X vector corresponding to A_1B_1 is .375 and that this is equal to

$$\overline{A_1B_1} - \bar{A}_1 - \bar{B}_1 + \bar{Y} = 6.00 - 5.250 - 6.500 + 6.125 = .375$$

Thus, this regression coefficient is an estimate of

$$\alpha_i\beta_j = \mu_{ij} - \mu_i - \mu_j + \mu$$

where $i = 1$ and $j = 1$.

12.6 Squared Multiple *R*'s

Because we have a squared multiple correlation with eleven subscripts, it will be convenient to simplify the notation for this correlation coefficient and also for the squared multiple correlations involving various subsets of the eleven X vectors. Thus, we let

$$A = \alpha_i \quad \text{be the subset } X_1, X_2$$

corresponding to the $a - 1 = 2X$ vectors for A,

$$B = \beta_j \quad \text{be the subset } X_3, X_4, X_5$$

corresponding to the $b - 1 = 3X$ vectors for B, and

$$C = \alpha_i\beta_j \quad \text{be the subset } X_6, X_7, X_8, X_9, X_{10}, X_{11}$$

corresponding to the $(a - 1)(b - 1) = 6X$ vectors that carry the $A \times B$ interaction.

Then we define

$$R^2_{Y.ABC} = R^2(\alpha_i, \beta_j, \alpha_i\beta_j) \tag{12.7}$$

as the squared multiple correlation coefficient of Y with all eleven X vectors;

$$R^2_{Y.BC} = R^2(\beta_j, \alpha_i\beta_j) \tag{12.8}$$

as the squared multiple correlation of Y with the nine X vectors for the levels of B *and* the $A \times B$ interactions;

$$R^2_{Y.AC} = R^2(\alpha_i, \alpha_i\beta_j) \tag{12.9}$$

as the squared multiple correlation of Y with the eight X vectors corresponding to the levels of A *and* the $A \times B$ interaction; and

$$R^2_{Y.AB} = R^2(\alpha_j, \beta_j) \tag{12.10}$$

as the squared multiple correlation of Y with the five X vectors corresponding to the levels of A *and* the levels of B.

The four squared multiple correlation coefficients as defined by (12.7), (12.8), (12.9), and (12.10) were obtained with three sequential runs through the computer and are given in Table 12.6. The degrees of freedom associated with each of these squared multiple correlation coefficients will be equal to the number of X vectors on which each is

TABLE 12.6 Values of R^2 and the values of $R^2_{Y(C.AB)}$, $R^2_{Y(A.BC)}$, and $R^2_{Y(B.AC)}$ obtained from the values of R^2

R	R^2	d.f.	SS_{reg}
$R_{Y.ABC}$	.85832	11	127.2192
$R_{Y.AB}$	.27154	5	40.2473
$R_{Y.BC}$	.73468	9	108.8934
$R_{Y.AC}$	.76671	8	113.6408
Semipartials			SS_{reg}
$R^2_{Y(C.AB)} = R^2_{Y.ABC} - R^2_{Y.AB} = .58678$			86.9718
$R^2_{Y(A.BC)} = R^2_{Y.ABC} - R^2_{Y.BC} = .12364$			18.3258
$R^2_{Y(B.AC)} = R^2_{Y.ABC} - R^2_{Y.AC} = .09161$			13.5783
		Σ	118.8759

based. For example, A has two X vectors, B has three X vectors, and C has six X vectors. Then $R^2_{Y.ABC}$ will have $2 + 3 + 6 = 11$ degrees of freedom. Similarly, $R^2_{Y.BC}$ will have $3 + 6 = 9$ degrees of freedom, $R^2_{Y.AC}$ will have $2 + 6 = 8$ d.f., and $R^2_{Y.AB}$ will have $2 + 3 = 5$ d.f.

12.7 Squared Semipartial R's and Tests of Significance

Given the values of the squared multiple correlation coefficients in Table 12.6, we can proceed to find the values of the squared semipartial multiple correlation coefficients* in which we are interested. For example,

$$R^2_{Y(C.AB)} = R^2_{Y.ABC} - R^2_{Y.AB} = .85832 - .27154 = .58678$$

is the proportion of the total sum of squares that can be accounted for by the $C = A \times B$ interaction over and above that already accounted for by X vectors corresponding to the combined A and B effects. The degrees of freedom associated with this semipartial correlation coefficient will be equal to $11 - 5 = 6 = (a - 1)(b - 1)$. As a test of significance we have

$$F = \frac{R^2_{Y(C.AB)}/(a - 1)(b - 1)}{(1 - R^2_{Y.ABC})/(n - ab)} \tag{12.11}$$

where $n = 32$ is the total number of observations and $ab = (3)(4) = 12$ is the total number of treatment combinations. Note that we have a

* We shall, hereafter, refer to a semipartial multiple correlation coefficient as a semipartial correlation coefficient. It will be obvious from the discussion and examples cited that they are semipartial *multiple* correlation coefficients.

total of $k = 11X$ vectors and that $n - k - 1 = 32 - 11 - 1 = 20 = n - ab$.

Substituting in (12.11) with the appropriate values given in Table 12.6, we have

$$F = \frac{.58678/6}{(1 - .85832)/20} = 13.8053$$

which is significant with 6 and 20 d.f. and with $\alpha = .05$. Thus, we can conclude that there is a significant interaction effect in our example. The interaction is highly disordinal, as indicated by the graph of the means for the levels of A against the levels of B in Figure 12.1.

In some experiments, depending on the nature of the factors, the tests of the main effects of A, averaged over the levels of B, and the main effects of B, averaged over the levels of A, in the presence of a significant $A \times B$ interaction may or may not be meaningful. In the interests of completeness, we proceed to show how these tests can be made (not that they necessarily should or should not be made in the presence of a significant $A \times B$ interaction).

Using the values given in Table 12.6, we have the squared semipartial correlation

$$R^2_{Y(A.BC)} = R^2_{Y.ABC} - R^2_{Y.BC} = .85832 - .73468 = .12364$$

with $11 - 3 - 6 = 2 = a - 1$ d.f. Given that we have already taken into account the B and $A \times B$ interaction, the additional proportion of the total sum of squares accounted for by A is .12364. Then we

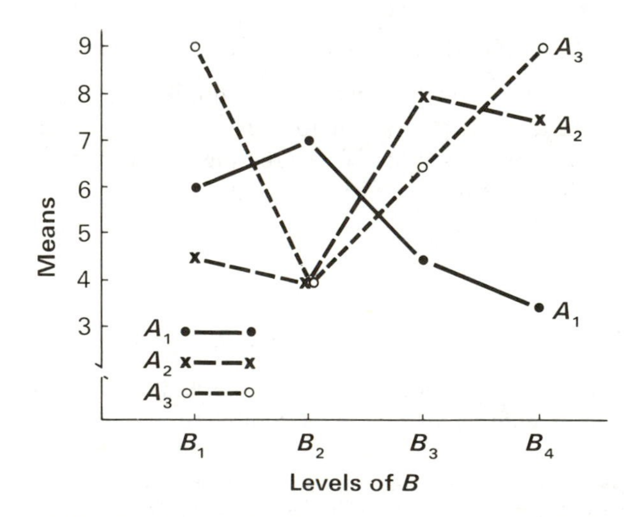

Figure 12.1 Graph of the $A \times B$ interaction for the data in Table 12.2.

have

$$F = \frac{.12364/2}{(1 - .85832)/20} = 8.7267$$

with 2 and 20 d.f.; this is also significant with $\alpha = .05$.

Finally, we have

$$R^2_{Y(B.AC)} = R^2_{Y.ABC} - R^2_{Y.AC} = .85832 - .76671 = .09161$$

with $11 - 2 - 6 = 3 = b - 1$ d.f. The proportion of the total sum of squares accounted for by B after taking into account the A and $A \times B$ interaction effects is equal to .09161. For the test of significance we have

$$F = \frac{.09161/3}{(1 - .85832)/20} = 4.3107$$

which is significant with 3 and 20 degrees of freedom and with $\alpha = .05$.

If we multiply each of the squared multiple correlation coefficients in Table 12.6 by $SS_{tot} = 148.2188$, we obtain the regression sum of squares associated with each of the coefficients. These values are also shown in Table 12.6. All of the tests of significance we have made with respect to the squared semipartial correlations can also be made with differences between the regression sums of squares. For example, the total regression sum of squares is 127.2192, and $127.2192 - 40.2473 = 86.9719$, which is equal, within rounding errors, to $R^2_{Y(C.AB)}SS_{tot} = (.58678) \times (148.2188) = 86.9718$. The value of 86.9719 is the increment in the regression sum of squares attributable to the $A \times B$ interaction after we have already taken into account the A and B effects. The residual or error sum of squares for testing this increment for significance is $SS_{res} = (1 - R^2_{Y.ABC})SS_{tot} = (1 - .85832)(148.2188) = 21.000$, rounded. It is also true that $SS_{res} = 21.000$ is equal to $SS_W = 21.000$ in the analysis of variance. Then, to determine whether the increment in the regression sum of squares attributable to the $A \times B$ interaction represents a significant increment, we have

$$F = \frac{86.9718/6}{21.00/20} = 13.8050$$

which is equal, within rounding errors, to the F test of $R^2_{Y(C.AB)}$.

Similar tests can be made with respect to the other regression sums of squares given in Table 12.6. For example, for the A effect we have

$$F = \frac{(127.2192 - 108.8934)/2}{21.00/20} = 8.7266$$

which is equal, within rounding errors, to the F test of $R^2_{Y(A.BC)}$.

The overall regression sum of squares is equal to 127.2192. Note, however, that the sum of the regression sums of squares is equal to 118.8759, which is not equal to the overall regression sum of squares. In other words, the regression sums of squares for A, B, and the $A \times B$ interaction do not represent an orthogonal partitioning of the overall regression sum of squares.

We believe that the F tests we have described for the data given in Table 12.2 are appropriate for experimental data, under the assumptions stated earlier in the chapter. In this respect we are in agreement with Carlson and Timm (1974) and Lewis and Keren (1977), who take much the same position with respect to a multiple regression analysis of the nonorthogonal two-way factorial experiment.

With an orthogonal design it will be true that

$$R^2_{Y.A} = R^2_{Y(A.B)} = R^2_{Y(A.C)} = R^2_{Y(A.BC)}$$
$$R^2_{Y.B} = R^2_{Y(B.A)} = R^2_{Y(B.C)} = R^2_{Y(B.AC)}$$
$$R^2_{Y.C} = R^2_{Y(C.A)} = R^2_{Y(C.B)} = R^2_{Y(C.AB)}$$

and the problem of how to determine the contribution of A, B, and $C = A \times B$ to the regression sum of squares does not exist. However, if the design is nonorthogonal, then we have six possible *hierarchical* analyses:

$$R^2_{Y.ABC} = R^2_{Y.A} + R^2_{Y(B.A)} + R^2_{Y(C.AB)}$$
$$R^2_{Y.ABC} = R^2_{Y.A} + R^2_{Y(C.A)} + R^2_{Y(B.AC)}$$
$$R^2_{Y.ABC} = R^2_{Y.B} + R^2_{Y(A.B)} + R^2_{Y(C.AB)}$$
$$R^2_{Y.ABC} = R^2_{Y.B} + R^2_{Y(C.B)} + R^2_{Y(A.BC)}$$
$$R^2_{Y.ABC} = R^2_{Y.C} + R^2_{Y(A.C)} + R^2_{Y(B.AC)}$$
$$R^2_{Y.ABC} = R^2_{Y.C} + R^2_{Y(B.C)} + R^2_{Y(A.BC)}$$

each of which would constitute an orthogonal partitioning of the overall regression sum of squares. But with unequal n's it will, in general, be true that

$$R^2_{Y.A} \neq R^2_{Y(A.B)} \neq R^2_{Y(A.C)} \neq R^2_{Y(A.BC)}$$

It will thus make a difference whether we test $R^2_{Y.A}$, $R^2_{Y(A.B)}$ $R^2_{Y(A.C)}$, or $R^2_{Y(A.BC)}$ for significance to determine if there is a significant A effect. And these same considerations apply to the test of significance of the B and the $C = A \times B$ effects. As we pointed out before, with a nonorthogonal design there is no *unique* way in which to partition the overall regression sum of squares into a sum of squares for the A, B, and $C = A \times B$ interaction.

TABLE 12.7 Population models and corresponding values of R^2

Population model	R^2
1. $\mu_{ij} = \mu$	0
2. $\mu_{ij} = \mu + \alpha_i$	$R^2_{Y.A}$
3. $\mu_{ij} = \mu + \beta_j$	$R^2_{Y.B}$
4. $\mu_{ij} = \mu + \alpha_i\beta_j$	$R^2_{Y.C}$
5. $\mu_{ij} = \mu + \alpha_i + \beta_j$	$R^2_{Y.AB}$
6. $\mu_{ij} = \mu + \alpha_i + \alpha_i\beta_j$	$R^2_{Y.AC}$
7. $\mu_{ij} = \mu + \beta_j + \alpha_i\beta_j$	$R^2_{Y.BC}$
8. $\mu_{ij} = \mu + \alpha_i + \beta_j + \alpha_i\beta_j$	$R^2_{Y.ABC}$

$A = \alpha_i \qquad B = \beta_j \qquad C = \alpha_i\beta_j$
$\Sigma\alpha_i = \Sigma\beta_j = \Sigma_i\alpha_i\beta_j = \Sigma_j\alpha_i\beta_j = 0$

12.8 Model Comparisons and Tests of Significance

Table 12.7 shows eight possible models for a two-factor fixed-effect design.* At the right we show the values of R^2 for each model. For clarity in subscript notation we let $A = \alpha_i$, $B = \beta_j$, and $C = \alpha_i\beta_j$. We also assume the following side conditions:

$$\Sigma\alpha_i = \Sigma\beta_j = \sum_i \alpha_i\beta_j = \sum_j \alpha_i\beta_j = 0$$

Now suppose that we want to test the null hypothesis that some parameter of interest, say α_i, is equal to zero. Then we need to compare a model that includes the parameter with another model that includes *all relevant parameters except the one of interest.* Throughout our discussion in this chapter we have assumed that *all* parameters are relevant; that is, we have assumed that Model 8 is the relevant population model. Then for the test of the null hypothesis $\alpha_i\beta_j = 0$ we compare Model 8 with Model 5, and for the test of significance we have

$$F = \frac{(R^2_{Y.ABC} - R^2_{Y.AB})/(a-1)(b-1)}{(1 - R^2_{Y.ABC})/(n - ab)} \tag{12.12}$$

* For a more detailed discussion of these models and their relevances to hypothesis testing, see Carlson and Timm (1974), Lewis and Keren (1977), and Herr and Gaebelein (1978).

For the test of the null hypothesis $\alpha_i = 0$, we compare Model 8 with Model 7, and for the test of significance we have

$$F = \frac{(R^2_{Y.ABC} - R^2_{Y.BC})/(a-1)}{(1 - R^2_{Y.ABC})/(n - ab)} \tag{12.13}$$

Similarly, for the test of the null hypothesis $\beta_j = 0$ we compare Model 8 with Model 6. And for the test of significance we have

$$F = \frac{(R^2_{Y.ABC} - R^2_{Y.AC})/(b-1)}{(1 - R^2_{Y.ABC})/(n - ab)} \tag{12.14}$$

There are some statisticians who recommend that if the test of the null hypothesis $\alpha_i\beta_j = 0$, as given by (12.12), is *not* rejected, then we should discard Model 8 in favor of Model 5 in which the $\alpha_i\beta_j$ component does not appear. The choice of Model 5, however, means that the experimenter is willing to assume that all $\alpha_i\beta_j$ are equal to zero. Under this assumption, the test of the null hypothesis $\alpha_i = 0$ would be a comparison between Model 5 and Model 3. For the test of significance we would then have

$$F = \frac{(R^2_{Y.AB} - R^2_{Y.B})/(a-1)}{(1 - R^2_{Y.AB})/(n - a - b + 1)} \tag{12.15}$$

Similarly, for the test of the null hypothesis that $\beta_j = 0$ the comparison would be between Model 5 and Model 2, and for the test of significance we would have

$$F = \frac{(R^2_{Y.AB} - R^2_{Y.A})/(b-1)}{(1 - R^2_{Y.AB})/(n - a - b + 1)} \tag{12.16}$$

If a two-factor experiment is orthogonal, that is, if an equal number of observations are obtained for each treatment combination, then, judging from published research, almost all experimenters analyze their data with Model 8 as the relevant population model. To do otherwise, would be to ignore the factorial structure of the experiment. But if Model 8 is the choice for an orthogonal experiment, why should it be otherwise for a nonorthogonal experiment, where the nonorthogonality is the result of randomly missing observations for some of the treatment combinations?

In some cases an experimenter may have strong prior reasons for believing that all $\alpha_i\beta_j$ are equal to zero. Thus, the experiment may be designed with $n = 1$ observation for each treatment combination. The obvious choice of a relevant population model, in this instance, is Model 5.

In general, however, we believe that for experimental data, Model 8 is the relevant population model, not only for orthogonal designs but also for nonorthogonal designs.

12.9 Unweighted Means Analysis of Variance

For experimental data, under the conditions described earlier in the chapter, an *unweighted means analysis of variance* is often a reasonable alternative to a multiple regression analysis. Assume, for the moment, that we have $n_{ij} = n$ for each A_iB_j treatment combination. Then, if we subtract $\bar{Y}$ from both sides of (12.4), square, and sum over all *abn* observations, we obtain

$$n\Sigma(\overline{A_iB_j} - \bar{Y})^2 = nb\Sigma(\bar{A}_i - \bar{Y})^2 + na\Sigma(\bar{B}_j - \bar{Y})^2 + n\Sigma(\overline{A_iB_j} - \bar{A}_i - \bar{B}_j + \bar{Y})^2 \qquad (12.17)$$

The term on the left is SS_T, the treatment sum of squares, and the three terms on the right correspond to SS_A, SS_B, and SS_{AB}, respectively, and, in the case of equal *n*'s, provide an orthogonal partitioning of SS_T.

If the *n*'s are unequal and we wish to use an unweighted means analysis of variance, the common *n* for each of the terms in (12.17) is replaced by the *harmonic mean*, $\tilde{n}$, of the number of observations for each A_iB_j treatment combination. The harmonic mean will be given by

$$\tilde{n} = \frac{ab}{\Sigma(1/n_{ij})} \qquad (12.18)$$

where a is the number of levels of A, b is the number of levels of B, and n_{ij} is the variable number of observations for each of the A_iB_j treatment combinations. The summation is over all *ab* values of $1/n_{ij}$. For the data in Table 12.2 we have

$$\tilde{n} = \frac{(3)(4)}{\frac{1}{3} + \frac{1}{3} + \frac{1}{2} + \cdots + \frac{1}{3}} = 2.526$$

rounded.

If we replace *n* by the harmonic mean, $\tilde{n}$, in (12.17), we have

$$\tilde{n}\Sigma(\overline{A_iB_j} - \bar{Y})^2 = \tilde{n}b\Sigma(\bar{A}_i - \bar{Y})^2 + \tilde{n}a\Sigma(\bar{B}_j - \bar{Y})^2 + \tilde{n}\Sigma(\overline{A_iB_j} - \bar{A}_i - \bar{B}_j + \bar{Y})^2$$

and the three terms on the right will provide an orthogonal partitioning of the "treatment sum of squares" defined by

$$\tilde{n}\Sigma(\overline{A_iB_j} - \bar{Y})^2$$

The values of $(\bar{A}_i - \bar{Y})$, $(\bar{B}_j - \bar{Y})$, and $(\overline{A_iB_j} - \bar{A}_i - \bar{B}_j + \bar{Y})$ are given in Table 12.5. Using these values, we obtain

$$\tilde{n}b\Sigma(\bar{A}_i - \bar{Y})^2 = (2.526)(4)[(-.875)^2 + (-.125)^2 + (1.000)^2] = 17.998$$

$$\tilde{n}a\Sigma(\bar{B}_j - \bar{Y}) = (2.526)(3)[(.375)^2 + (-1.125)^2 + \cdots + (.542)^2] = 13.208$$

and

$$\tilde{n}\Sigma(\overline{A_iB_j} - \bar{A}_i - \bar{B}_j + \bar{Y})^2 = (2.526)[(.375)^2 + (2.875)^2 + \cdots + (1.333)^2] = 80.094$$

The within treatment sum of squares is calculated in the usual way and for the data in Table 12.2 is equal to 21.00 with $32 - 12 = 20$ degrees of freedom. Thus, we have $MS_W = 21.00/20 = 1.05$. For tests of significance we have

$$F = \frac{MS_A}{MS_W} = \frac{17.998/2}{1.05} = 8.570$$

$$F = \frac{MS_B}{MS_W} = \frac{13.208/3}{1.05} = 4.233$$

and

$$F = \frac{MS_{AB}}{MS_W} = \frac{80.094/6}{1.05} = 12.713$$

and all values of F are significant with $\alpha = .05$.

If you compare the sums of squares for A, B, and $A \times B$ obtained with an unweighted means analysis of variance with those based on semipartial correlations, which are given in Table 12.6, you will observe that the two methods of analysis yield fairly comparable results. This happens to be true for the data in Table 12.2, but it will not necessarily be true for other data sets.

Exercises

12.1 A two-factor experiment with $a = 2$ levels of A and $b = 3$ levels of B was planned with $n = 5$ subjects assigned at random to each treatment combination. For some of the treatment combinations there was a loss of subjects. We assume that the missing observations are random and in no way related to the nature of the treatments themselves.

A_1			A_2		
B_1	B_2	B_3	B_1	B_2	B_3
1	6	3	5	6	4
2	7	4	6	7	5
3	8	5	7	8	6
4		6		9	
5		7		10	

(a) Calculate SS_T and SS_W.
(b) Is $F = MS_T/MS_W$ significant with $\alpha = .05$?

12.2 Effect coding was used with the data in Exercise 12.1. X_1 was coded for the levels of A, and X_2 and X_3 for the levels of B. The interaction vectors were $X_4 = X_1X_2$ and $X_5 = X_1X_3$. Computer output gave the following values: $R^2_{Y.12345} = .66615$, $R^2_{Y.123} = .58501$, $R^2_{Y.145} = .30680$, and $R^2_{Y.2345} = .57342$.
(a) What does $R^2_{Y.12345} - R^2_{Y.123}$ measure?
(b) Test $R^2_{Y.12345} - R^2_{Y.123}$ for significance.
(c) What does $R^2_{Y.12345} - R^2_{Y.2345}$ measure?
(d) Test $R^2_{Y.12345} - R^2_{Y.2345}$ for significance.
(e) What does $R^2_{Y.12345} - R^2_{Y.145}$ measure?
(f) Test $R^2_{Y.12345} - R^2_{Y.145}$ for significance.
(g) What are the values of b_1, b_2, b_3, b_4, and b_5, and what are they estimates of?
(h) Is $SS_{tot}(1 - R^2_{Y.12345})$ equal to SS_W in the analysis of variance?
(i) Is $SS_{tot}R^2_{Y.12345}$ equal to SS_T in the analysis of variance?
(j) Calculate the regression sum of squares for A, B, and $A \times B$.

12.3 Analyze the data in Exercise 12.1 using an unweighted means analysis of variance. Compare the sums of squares for A, B, and $A \times B$ and the tests of significance with those of the multiple regression analysis.

12.4 We have a nonorthogonal two-factor experiment.
(a) If $R^2_{Y.AB} - R^2_{Y.A}$ is tested for significance, what is the denominator for the F ratio?
(b) What is the relevant population model, and what assumptions does it involve?

12.5 Which of the eight population models discussed in the chapter involves the fewest assumptions? Explain why.

12.6 Explain each of the following concepts:

effect coding

harmonic mean

nonorthogonal design

unweighted means analysis of variance

13

The Nonorthogonal 2^k Factorial Experiment

13.1 Introduction

In the preceding chapter we compared the results obtained with a multiple regression analysis and an unweighted means analysis of variance for a nonorthogonal 3×4 factorial experiment. Although the sums of squares we obtained with both analyses were comparable, they were not equivalent. Consequently, the F tests of significance of the A, B, and $A \times B$ effects with the regression analysis and the unweighted means analysis were not equivalent. We also pointed out that the two methods of analysis would, in general, not result in equivalent tests of significance. What we failed to emphasize, however, is that the nonequivalence of the two methods of analysis applies only to those nonorthogonal designs in which the factors have more than two levels.

In this chapter we consider a nonorthogonal design in which the two factors, A and B, both have two levels. As we shall see, if both A and B have two levels, then a regression analysis and an unweighted means analysis will result in equivalent sums of squares and tests of significance.

13.2 An Unweighted Means Analysis of the Nonorthogonal 2^2 Factorial Experiment

Table 13.1 shows the values of Y for a factorial experiment in which both A and B have two levels. For the total sum of squares, we have

$$SS_{tot} = (1)^2 + (2)^2 + \cdots + (5)^2 - \frac{(53)^2}{14} = 64.357$$

The sums for the four treatment combinations are 6, 16, 15, and 16 with n's of 3, 2, 5, and 4, respectively. Then the *weighted* treatment sum of squares will be

$$SS_T = \frac{(6)^2}{3} + \frac{(16)^2}{2} + \frac{(15)^2}{5} + \frac{(16)^2}{4} - \frac{(53)^2}{14} = 48.357$$

and the within treatment sum of squares can be obtained by subtraction. Thus, we have

$$SS_W = SS_{tot} - SS_T = 64.357 - 48.357 = 16.000$$

Table 13.2 shows the means for each of the treatment combinations and the *unweighted* means: $\bar{A}_1 = 5.00$, $\bar{A}_2 = 3.50$, $\bar{B}_1 = 2.50$, $\bar{B}_2 = 6.00$, and $\bar{Y} = 4.25$. The n's for the four treatment combinations are 3, 2, 5, and 4. Then for the harmonic mean we have

$$\tilde{n} = \frac{(2)(2)}{\frac{1}{3} + \frac{1}{2} + \frac{1}{5} + \frac{1}{4}} = 3.117$$

TABLE 13.1 Values of Y for a 2×2 factorial experiment with unequal n's. The means for each treatment combination are the same as those shown in Table 13.2.

Treatment combination	Values of Y	n	$\bar{Y}_{ij}$
A_1B_1	1, 2, 3	3	2.00
A_1B_2	7, 9	2	8.00
A_2B_1	2, 1, 4, 5, 3	5	3.00
A_2B_2	4, 3, 4, 5	4	4.00

TABLE 13.2 Treatment means and the unweighted means $\bar{A}_i$, $\bar{B}_j$, and $\bar{Y}$

	B_1	B_2	Means
A_1	2.00	8.00	5.00
A_2	3.00	4.00	3.50
Means	2.50	6.00	4.25

With an unweighted means analysis of variance, we have

$$SS_A = \tilde{n}b\Sigma(\bar{A}_i - \bar{Y})^2 = (3.117)(2)[(.75)^2 + (-.75)^2] = 7.01$$
$$SS_B = \tilde{n}a\Sigma(\bar{B}_j - \bar{Y})^2 = (3.117)(2)[(-1.75)^2 + (1.75)^2] = 38.18$$

and

$$SS_{AB} = \tilde{n}(\overline{A_iB_j} - \bar{A}_i - \bar{B}_j + \bar{Y})^2$$

For the means in Table 13.2, we have

$$(\overline{A_1B_1} - \bar{A}_1 - \bar{B}_1 + \bar{Y})^2 = (2.00 - 5.00 - 2.50 + 4.25)^2 = 1.75$$
$$(\overline{A_1B_2} - \bar{A}_1 - \bar{B}_2 + \bar{Y})^2 = (8.00 - 5.00 - 6.00 + 4.25)^2 = 1.75$$
$$(\overline{A_2B_1} - \bar{A}_2 - \bar{B}_1 + \bar{Y})^2 = (3.00 - 3.50 - 2.50 + 4.25)^2 = 1.75$$
$$(\overline{A_2B_2} - \bar{A}_2 - \bar{B}_2 + \bar{Y})^2 = (4.00 - 3.50 - 6.00 + 4.25)^2 = 1.75$$

Then

$$\tilde{n}\Sigma(\overline{A_iB_j} - \bar{A}_i - \bar{B}_j + \bar{Y})^2 = (3.117)(4)(1.75) = 19.48$$

Because SS_A, SS_B, and SS_{AB} each have 1 d.f., each will be a mean square. We have $SS_W = 16.00$ with 10 d.f. and $MS_W = 16.00/10 = 1.60$. Then for the tests of significance we have

$$F = \frac{MS_A}{MS_W} = \frac{7.01}{1.60} = 4.38$$

$$F = \frac{MS_B}{MS_W} = \frac{38.18}{1.60} = 23.86$$

and

$$F = \frac{MS_{AB}}{MS_W} = \frac{19.48}{1.60} = 12.17$$

With $\alpha = .05$ and with 1 and 10 d.f., MS_B and MS_{AB} are significant and MS_A is not.

TABLE 13.3 Effect coding for a 2 × 2 factorial experiment

	X_1	X_2	X_3
	A	B	$A \times B$
A_1B_1	1	1	1
A_1B_2	1	−1	−1
A_2B_1	−1	1	−1
A_2B_2	−1	−1	1

13.3 A Regression Analysis of the Nonorthogonal 2^2 Factorial Experiment

We now analyze the same data using a multiple regression analysis. Table 13.3 shows effect coding for a 2 × 2 factorial experiment. The first vector X_1 codes for the A effect, the second vector X_2 codes for the B effect, and the third vector $X_3 = X_1X_2$ is a product vector which carries the interaction effect. Table 13.4 shows the values of the X vectors associated with each of the Y values in the experiment.

Then, with the three X vectors shown in Table 13.4, we have

$$R^2_{Y.123} = .75138$$
$$R^2_{Y.12} = .44869$$
$$R^2_{Y.23} = .64242$$
$$R^2_{Y.13} = .15810$$

TABLE 13.4 Effect coding for the data in Table 13.1

X_1	X_2	X_3	Y
1	1	1	1
1	1	1	2
1	1	1	3
1	−1	−1	7
1	−1	−1	9
−1	1	−1	2
−1	1	−1	1
−1	1	−1	4
−1	1	−1	5
−1	1	−1	3
−1	−1	1	4
−1	−1	1	3
−1	−1	1	4
−1	−1	1	5

For the A effect, coded by X_1, we have the squared semipartial correlation coefficient

$$r^2_{Y(1.23)} = R^2_{Y.123} - R^2_{Y.23} = .75138 - .64242 = .10896$$

To test $r^2_{Y(1.23)}$ for significance, we have

$$F = \frac{R^2_{Y.123} - R^2_{Y.12}}{(1 - R^2_{Y.123})/(n - k - 1)} = \frac{.10896}{(1 - .75138)/10} = 4.38$$

and we observe that this F ratio is equal to the one we obtained in testing the A effect for significance with the unweighted means analysis of variance.

For the B effect, coded by X_2, we have the squared semipartial correlation coefficient

$$r^2_{Y(2.13)} = R^2_{Y.123} - R^2_{Y.13} = .75138 - .15810 = .59328$$

To test $r^2_{Y(2.13)}$ for significance we have

$$F = \frac{.59328}{.02486} = 23.86$$

where $.02486 = (1 - R^2_{Y.123})/(n - k - 1)$. We also observe that this F ratio is equal to the F obtained in the test of significance of the B effect in the unweighted means analysis of variance.

Similarly, for the $A \times B$ interaction, coded by X_3, we have the squared semipartial correlation coefficient

$$r^2_{Y(3.12)} = R^2_{Y.123} - R^2_{Y.12} = .75138 - .44869 = .30269$$

and for the test of significance of $r^2_{Y(3.12)}$ we have

$$F = \frac{.30269}{.02486} = 12.18$$

which is equal, within rounding errors, to the F test of the $A \times B$ interaction in the unweighted means analysis of variance.

13.4 Sums of Squares with the Regression Analysis

The regression sums of squares for the A, B, and $A \times B$ interaction effects are

$$SS_A = SS_{tot} r^2_{Y(1.23)} = (64.357)(.10896) = 7.01$$
$$SS_B = SS_{tot} r^2_{Y(2.13)} = (64.357)(.59328) = 38.18$$

and

$$SS_{AB} = SS_{tot} r^2_{Y(3.12)} = (64.357)(.30269) = 19.48$$

We observe that these sums of squares are the same as those we obtained with an *unweighted* means analysis of variance. For the sum of these three sums of squares, we have

$$SS_A + SS_B + SS_{AB} = 7.01 + 38.18 + 19.48 = 64.67$$

which is equal to the *unweighted* treatment sum of squares or

$$\tilde{n}\Sigma(\overline{A_iB_j} - \bar{Y})^2 \tag{13.1}$$

where $\bar{Y}$ is the unweighted mean of the four treatment means. In our example, we have

$$\bar{Y} = \frac{2.00 + 8.00 + 3.00 + 4.00}{4} = 4.25$$

we have

$$(\overline{A_1B_1} - \bar{Y})^2 = (2.00 - 4.25)^2 = 5.0625$$
$$(\overline{A_1B_2} - \bar{Y})^2 = (8.00 - 4.25)^2 = 14.0625$$
$$(\overline{A_2B_1} - \bar{Y})^2 = (3.00 - 4.25)^2 = 1.5625$$
$$(\overline{A_2B_2} - \bar{Y})^2 = (4.00 - 4.25)^2 = .0625$$

and for the sum of these squared deviations we have

$$\Sigma(\overline{A_iB_j} - \bar{Y})^2 = 20.75$$

Then for the unweighted treatment sum of squares as given by (13.1), we have

$$\tilde{n}\Sigma(\overline{A_iB_j} - \bar{Y})^2 = (3.117)(20.75) = 64.68$$

which is equal, within rounding errors, to the sum of the three regression sums of squares.

Earlier in the chapter we calculated the *weighted* treatment sum of squares and found that it was equal to 48.357. The overall regression sum of squares will, as we know, be given by $SS_{tot}R^2_{Y.123}$ and, in our example, we have

$$SS_{reg} = (64.357)(.75138) = 48.357$$

and SS_{reg} is equal to the *weighted* treatment sum of squares.

13.5 The General Case of the 2^k Nonorthogonal Factorial Experiment

Although we have shown the equivalence between a regression analysis and an unweighted means analysis of variance for a nonorthogonal 2×2 factorial experiment, it can be shown that this equivalence holds

TABLE 13.5 Effect coding for a 2^3 factorial experiment

	X_1	X_2	X_3	X_4	X_5	X_6	X_7
	A	B	C	$A \times B$	$A \times C$	$B \times C$	$A \times B \times C$
$A_1B_1C_1$	1	1	1	1	1	1	1
$A_1B_1C_2$	1	1	−1	1	−1	−1	−1
$A_1B_2C_1$	1	−1	1	−1	1	−1	−1
$A_1B_2C_2$	1	−1	−1	−1	−1	1	1
$A_2B_1C_1$	−1	1	1	−1	−1	1	−1
$A_2B_1C_2$	−1	1	−1	−1	1	−1	1
$A_2B_2C_1$	−1	−1	1	1	−1	−1	1
$A_2B_2C_2$	−1	−1	−1	1	1	1	−1

no matter how many factors we have, provided that each of the factors has only two levels.*

For example if we have $k = 3$ factors, each with two levels, then we will have $2^3 = 8$ treatment combinations. We will thus need seven X vectors to code for the A, B, C, $A \times B$, $A \times C$, $B \times C$, and $A \times B \times C$ effects. Table 13.5 shows effect coding for the 2^3 factorial experiments.

If the 2^3 factorial experiment is nonorthogonal, it will still be true that with a regression analysis the residual sum of squares will be equal to the within treatment sum of squares or

$$SS_{tot}(1 - R^2_{Y.1234567}) = SS_W$$

It will also be true that the overall regression sum of squares will be equal to the *weighted* treatment sum of squares or

$$SS_{tot}R^2_{Y.1234567} = \Sigma n_{ij}(\overline{A_iB_j} - \bar{Y}..)^2$$

where n_{ij} is the number of observations on which the mean $\overline{A_iB_j}$ is based and $\bar{Y}..$ is the weighted mean of the treatment means or, more simply,

$$\bar{Y}.. = \frac{\Sigma Y}{\Sigma n_{ij}} \tag{13.2}$$

The regression sum of squares for the A effect, coded by X_1, will be given by

$$SS_{tot}(R^2_{Y.1234567} - R^2_{Y.234567}) = SS_A$$

and this sum of squares will be equal, within rounding errors, to the sum of squares for A with an unweighted means analysis of variance.†

* See, for example, Horst and Edwards (1981, 1982).

† It has been shown by Speed and Monlezun (1979) that for the nonorthogonal 2^k factorial experiment, all sums of squares with an unweighted means analysis of variance are χ^2 with 1 d.f. and that, consequently, all tests of significance are exact.

Similarly, the regression sum of squares for the B effect, coded by X_2, will be given by

$$SS_{tot}(R^2_{Y.1234567} - R^2_{Y.134567}) = SS_B$$

and this sum of squares will be equal, within rounding errors, to the sum of squares for the B effect with an unweighted means analysis of variance.

In the same manner, we can obtain the regression sums of squares for each of the other effects. If we do so and then add these sums of squares, their sum will be equal to the unweighted treatment sum of squares in the unweighted means analysis of variance.

If the design is orthogonal, that is, if we have an equal number of observations for each treatment combination, then, as can been seen in Table 13.5, all X vectors will be orthogonal. In this case, the harmonic mean $\tilde{n}$ will simply be equal to the number of observations on which each treatment mean is based. Consequently, with an orthogonal design, the weighted treatment sum of squares will be equal to the unweighted treatment sum of squares. With an orthogonal design, it will also be true that the regression sums of squares for each effect will add up to the overall regression sum of squares.

Exercises

13.1 We have a 2×2 factorial experiment with the following values of Y:

A_1B_1	A_1B_2	A_2B_1	A_2B_2
1	3	5	9
3	5	7	11

(a) Use effect coding and calculate r^2_{Y1}, r^2_{Y2}, and r^2_{Y3}.
(b) Calculate $R^2_{Y.123}$.
(c) Show that $SS_{tot}R^2_{Y.123} = SS_T$ and that $SS_{tot}(1 - R^2_{Y.123}) = SS_W$.
(d) Show that $SS_{tot}r^2_{Y1} = SS_A$, $SS_{tot}r^2_{Y2} = SS_B$, and $SS_{tot}r^2_{Y3} = SS_{AB}$.

13.2 We have a nonorthogonal 2^2 factorial experiment.

(a) We calculate the weighted treatment sum of squares. What sum of squares will this be equal to in the regression analysis?

(b) We calculate the within treatment sum of squares. What sum of squares will this be equal to in the regression analysis?

(c) We calculate the unweighted treatment sum of squares. What sum of squares will this be equal to in the regression analysis?

(d) We calculate $SS_{tot}(1 - R^2_{Y.123})$. What sum of squares will this be equal to in the analysis of variance?

(e) We calculate $SS_{tot}R^2_{Y.123}$. What sum of squares will this be equal to in the analysis of variance?

(f) We calculate $SS_{tot}(R^2_{Y.123} - R^2_{Y.12})$. What sum of squares will this be equal to in the analysis of variance?

13.3 We have an orthogonal 2^3 factorial experiment with n observations for each treatment combination. Show that the harmonic mean $\tilde{n}$ will be equal to n.

14

The Analysis of Covariance for a Completely Randomized Design

14.1 Introduction

An experimenter divides subjects at random into two groups.* One group is to receive Treatment 1 and the other Treatment 2. Prior to being administered the treatments, each subject in each group is tested

* It is not necessary that there be two groups; we may have any number of groups or treatments. Nor is it necessary that we have the same number of subjects in each of the groups, as we do in the example to be described.

TABLE 14.1 Values of a covariate (X) and a dependent variable (Y) in the absence of treatment effects

	T_1		T_2	
	X	Y	X	Y
	5	6	2	1
	3	4	6	7
	1	0	4	3
	4	3	7	8
	6	7	3	2
Σ	19	20	22	21

under a standard condition on the dependent variable Y. We shall refer to these initial or pretest measures as a *covariate* and designate them by X.* After the treatments are administered, each subject is again tested on the dependent variable. We shall refer to these posttest measures as Y. It is the difference between the means of the two groups on the posttest measures that is of experimental interest. The method for analyzing the data from an experiment in which one or more covariates are available in addition to the dependent variable Y is commonly referred to as the *analysis of covariance.*

Table 14.1 gives the values of the covariate X and the dependent variable Y for each of $n_i = 5$ subjects in each treatment group. The Y values shown in Table 14.1 are those that we might expect to obtain in the absence of any real treatment effects and when the measure of the dependent variable is highly reliable. Figure 14.1 shows a plot of the Y measures against the X measures; it is clear that the correlation between the X and Y measures is quite high and positive.

Now suppose that Treatment 1 does indeed have an influence on the dependent variable and that the influence is additive. We shall assume, for example, that the influence of Treatment 1 is such that it increases the values of Y given in Table 14.1 by 5 points. At the same time we shall assume that Treatment 2 has had no effect and that the Y measures for Treatment 2 remain unchanged. Table 14.2 gives the values of the covariate X and the values of Y for each subject under these assumptions. Figure 14.2 shows a plot of the Y measures against the X measures for each group of subjects. The two lines drawn in the figure are the regression lines of Y on X as obtained separately for

* The covariate need not be the same as the dependent variable. It can be any variable that we have reason to believe will be correlated with the dependent variable.

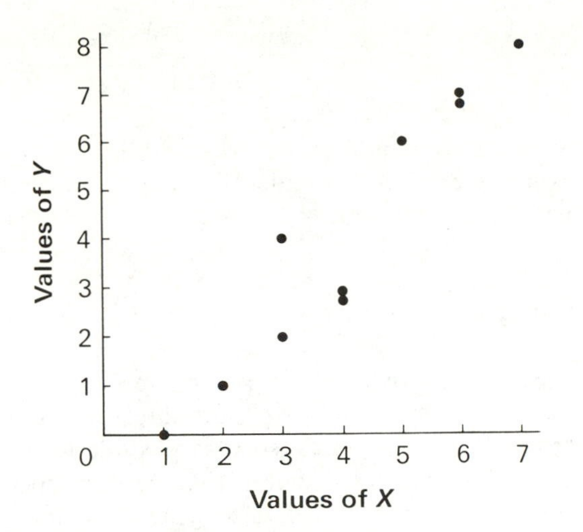

Figure 14.1 Plot of the Y values against the X values for the data in Table 14.1.

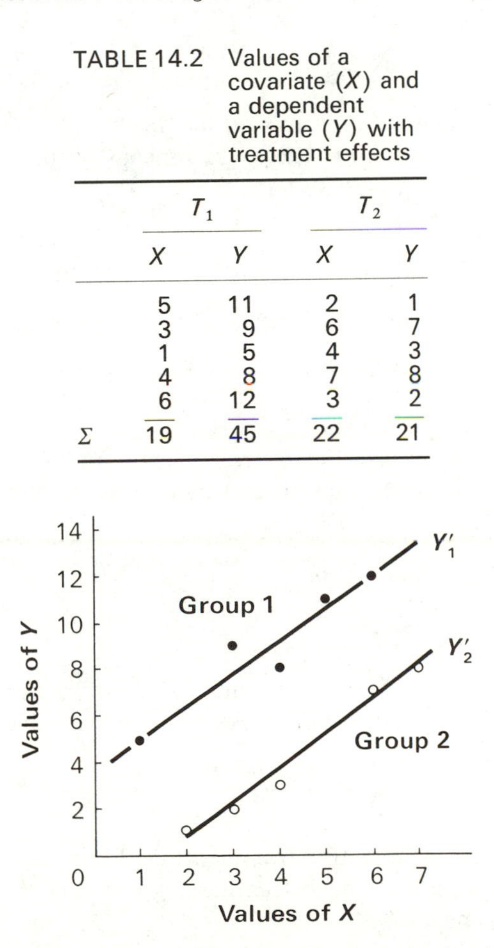

TABLE 14.2 Values of a covariate (X) and a dependent variable (Y) with treatment effects

	T_1		T_2	
	X	Y	X	Y
	5	11	2	1
	3	9	6	7
	1	5	4	3
	4	8	7	8
	6	12	3	2
Σ	19	45	22	21

Figure 14.2 Plot of the Y values against the X values for the data in Table 14.2. The lines shown are the regression lines for the two groups.

each group. As we would expect, the addition of a constant to each of the Y measures obtained under Treatment 1 has served to increase the value of the Y-intercept.

14.2 Partitioning the Total Product Sum

Just as we have shown that it is possible to partition the total sum of squares in a randomized group design into the within treatment and the treatment sum of squares, so it is possible to partition the total product sum (SP_{tot}) into the within treatment product sum (SP_W) and the treatment product sum (SP_T). In general, with k groups the total product sum will be given by

$$SP_{tot} = \Sigma(X - \bar{X})(Y - \bar{Y}) = \Sigma XY - \frac{(\Sigma X)(\Sigma Y)}{n} = \Sigma xy_{tot} \quad (14.1)$$

where the summation is over all $n = n_1 + n_2 + \cdots + n_k$ observations, and where $\bar{X}$ and $\bar{Y}$ are the overall means of the n observations. For the data in Table 14.2 we have

$$SP_{tot} = (5)(11) + (3)(9) + \cdots + (3)(2) - \frac{(41)(66)}{10} = 38.4 = \Sigma xy_{tot}$$

The product sum for treatments will be given by

$$SP_T = n_1(\bar{X}_1 - \bar{X})(\bar{Y}_1 - \bar{Y}) + n_2(\bar{X}_2 - \bar{X})(\bar{Y}_2 - \bar{Y}) + \cdots + n_k(\bar{X}_k - \bar{X})(\bar{Y}_k - \bar{Y}) = \Sigma xy_t \quad (14.2)$$

where $\bar{X}$ and $\bar{Y}$ are the overall means, and where $\bar{X}_i$ and $\bar{Y}_i$ are the means of the various treatment groups. An algebraic identity for (14.2) is

$$SP_T = \frac{(\Sigma X_1)(\Sigma Y_1)}{n_1} + \frac{(\Sigma X_2)(\Sigma Y_2)}{n_2} + \cdots + \frac{(\Sigma X_k)(\Sigma Y_k)}{n_k} - \frac{(\Sigma X)(\Sigma Y)}{n} = \Sigma xy_t \quad (14.3)$$

where ΣX_i and ΣY_i are the sums for the treatment groups and ΣX and ΣY are the overall sums. For the data in Table 14.2 we have

$$SP_T = \frac{(19)(45)}{5} + \frac{(22)(21)}{5} - \frac{(41)(66)}{10} = -7.20 = \Sigma xy_t$$

The product sum within treatments could be obtained by calculating the product sum for each treatment group and then summing these product sums; that is,

$$SP_W = SP_1 + SP_2 + \cdots + SP_k = \Sigma xy_1 + \Sigma xy_2 + \cdots + \Sigma xy_k = \Sigma xy_w \quad (14.4)$$

TABLE 14.3 Sums of squares and product sums for the data in Table 14.2

	Σx^2	Σy^2	Σxy
Treatment 1	14.80	30.00	20.00
Treatment 2	17.20	38.80	25.60
Within	32.00	68.80	45.60
Treatments	.90	57.60	−7.20
Total	32.90	126.40	38.40

For example, for Treatment 1 we would have

$$SP_1 = \Sigma(X_1 - \bar{X}_1)(Y_1 - \bar{Y}_1) = \Sigma xy_1$$

where the summation is over the n_1 observations for Treatment 1 and where $\bar{X}_1$ and $\bar{Y}_1$ are the X and Y means for Treatment 1. Because it is true that

$$SP_{tot} = SP_W + SP_T \quad \text{or} \quad \Sigma xy_{tot} = \Sigma xy_w + \Sigma xy_t$$

we can also obtain the within treatment product sum by subtraction. Thus, we also have

$$SP_W = SP_{tot} - SP_T \quad \text{or} \quad \Sigma xy_w = \Sigma xy_{tot} - \Sigma xy_t \tag{14.5}$$

and for the data in Table 14.2 we have

$$SP_W = 38.4 - (-7.2) = 45.6 = \Sigma xy_w$$

For the data in Table 14.2 we have partitioned the total sum of squares for the covariate X into the within treatment and the treatment sum of squares. We have also partitioned the total sum of squares for the Y variable into the within treatment and the treatment sum of squares. These calculations should by this time be familiar, so they are not shown but instead are summarized in Table 14.3, which also shows the partitioning of the total product sum.

14.3 The Residual Sum of Squares Using the Regression Coefficients for Each Treatment

Using the data shown in Table 14.3, we can find the residual sum of squares for Treatment 1 and Treatment 2. For example, we have

$$SS_{res_1} = \Sigma y_1^2 - \frac{(\Sigma xy_1)^2}{\Sigma x_1^2} = 30.0 - \frac{(20.0)^2}{14.8} = 2.973$$

and

$$SS_{res_2} = \Sigma y_2^2 - \frac{(\Sigma xy_2)^2}{\Sigma x_2^2} = 38.8 - \frac{(25.6)^2}{17.2} = .698$$

These two residual sums of squares are based on the two separate regression coefficients, $b_1 = \Sigma xy_1/\Sigma x_1^2 = 20.00/14.80 = 1.35$ and $b_2 = \Sigma xy_2/\Sigma x_2^2 = 25.60/17.20 = 1.49$, and, consequently, SS_{res_1} is at a minimum and so is SS_{res_2}. This means that the sum of squared deviations from the regression line Y_1' is at a minimum for the observations obtained in Treatment 1, and, similarly, the sum of squared deviations from the regression line Y_2' is at a minimum for the observations obtained in Treatment 2. We let the sum of these two residual sums of squares be SS_1. In general, with k groups we would have

$$SS_1 = SS_{res_1} + SS_{res_2} + \cdots + SS_{res_k} \tag{14.6}$$

and SS_1 will have, with $n_1 = n_2 = \cdots = n_i$, $k(n_i - 2)$ degrees of freedom. In our example, with $n_i = 5$ and $k = 2$, SS_1 has $(2)(5 - 2) = 6$ d.f. We also observe that an algebraic identity for (14.6) is

$$SS_1 = \Sigma y_w^2 - \left[\frac{(\Sigma xy_1)^2}{\Sigma x_1^2} + \frac{(\Sigma xy_2)^2}{\Sigma x_2^2} + \cdots + \frac{(\Sigma xy_k)^2}{\Sigma x_k^2}\right] \tag{14.7}$$

In our example, we have

$$SS_1 = 68.80 - \left[\frac{(20.0)^2}{14.10} + \frac{(25.60)^2}{17.20}\right] = 3.671$$

which is equal to

$$SS_{res_1} + SS_{res_2} = 2.973 + .698 = 3.671$$

14.4 The Residual Sum of Squares Using a Common Regression Coefficient

With k groups we will have k independent values of b_i. If we assume that these values of b_i are all estimates of the same common population value β, then the best estimate of this parameter, which we designate by b_w, will be

$$b_w = \frac{\Sigma xy_1 + \Sigma xy_2 + \cdots + \Sigma xy_k}{\Sigma x_1^2 + \Sigma x_2^2 + \cdots + \Sigma x_k^2} \tag{14.8}$$

The numerator of (14.8) is the product sum within treatments, and the denominator is the within treatment sum of squares for the covariate X. Then we also have

$$b_w = \frac{SP_W}{SS_{W(X)}} = \frac{\Sigma xy_w}{\Sigma x_w^2} \tag{14.9}$$

We can now obtain a residual sum of squares for each of the treatment groups from a common regression line with slope equal to $b_w =$

$\Sigma xy_w/\Sigma x_w^2 = 45.60/32.00 = 1.43$. We let this pooled residual sum of squares be SS_2, and it will be given by

$$SS_2 = \Sigma y_w^2 - \frac{(\Sigma xy_w)^2}{\Sigma x_w^2} \tag{14.10}$$

with $k(n_i - 1) - 1$ degrees of freedom. For the data in Table 14.3 we have

$$SS_2 = 68.8 - \frac{(45.6)^2}{32.0} = 3.82$$

with $2(5 - 1) - 1 = 7$ degrees of freedom.

14.5 Test for Homogeneity of the Regression Coefficients

Now SS_2 can never be smaller than SS_1, because SS_1 is based on the least squares value of b_i so as to minimize the residual sum of squares within each of the k treatments. If $b_1 = b_2 = \cdots = b_k = b_w$, then SS_2 and SS_1 will be equal, but if the values of b_i differ, then SS_2 will be larger than SS_1. We let

$$SS_3 = SS_2 - SS_1 \tag{14.11}$$

and, in our example, we have

$$SS_3 = 3.820 - 3.671 = .149$$

with $[k(n_i - 1) - 1] - [k(n_i - 2)] = k - 1$ degrees of freedom.

To determine whether the k regression coefficients differ significantly, we have

$$F = \frac{SS_3/(k - 1)}{SS_1/k(n_i - 2)} \tag{14.12}$$

and, in our example, with $k = 2$ and $n_i = 5$, we have

$$F = \frac{.149/1}{3.671/6} = .2435$$

a nonsignificant value with $\alpha = .05$.*

* Note that the test we have described for the difference between b_1 and b_2 is a general one. If we have k independent samples with measures of any two variables, X and Y, for each individual in each sample, we can calculate the regression coefficient b_i for each sample. Then the test we have described can be used to determine whether there are significant differences in the set of k regression coefficients. It is not necessary that we have an equal number of observations in each sample.

14.6 Test for Treatment Effects

Given that the two regression coefficients, $b_1 = 1.35$ and $b_2 = 1.49$, are homogeneous, we calculate

$$SS_4 = \Sigma y^2_{tot} - \frac{(\Sigma xy_{tot})^2}{\Sigma x^2_{tot}} \tag{14.13}$$

with $kn_i - 2 = 8$ degrees of freedom. Substituting in (14.13) with the appropriate values given in Table 14.3, we have

$$SS_4 = 126.4 - \frac{(38.4)^2}{32.9} = 81.58$$

We let the difference between SS_4 and SS_2 be SS_5, or

$$SS_5 = SS_4 - SS_2 \tag{14.14}$$

and SS_5 will have $[kn_i - 2] - [k(n_i - 1) - 1] = k - 1$ degrees of freedom. In our example we have

$$SS_5 = 81.581 - 3.820 = 77.761$$

with $k - 1 = 1$ d.f.

Under the various assumptions we have made,

$$F = \frac{SS_5/(k - 1)}{SS_2/(n - k - 1)} \tag{14.15}$$

will be a test to determine whether or not the treatment means differ significantly. In our example we have

$$F = \frac{77.761/1}{3.820/7} = 142.494$$

with 1 and 7 degrees of freedom. This is a significant value of F with $\alpha = .05$.

With a completely randomized design and without the measures of the covariate X, we would have had a simple analysis of variance for the Y measures. In this instance we would have had $MS_T = 57.6$ and $MS_W = 68.8/8 = 8.60$, and

$$F = \frac{MS_T}{MS_W} = \frac{57.6}{8.6} = 6.698$$

with 1 and 8 degrees of freedom. Two things have happened in the analysis of covariance: The error sum of squares with the covariance analysis (3.820) is a considerable reduction of the error sum of squares in the analysis of variance (68.8), and SS_5 in the analysis of covariance

(77.76) is considerably larger than the treatment sum of squares in the analysis of variance (57.6). Let us see why this is so.

The squared correlation coefficient between X and Y within treatments is

$$r_w^2 = \frac{(\Sigma xy_w)^2}{(\Sigma x_w^2)(\Sigma y_w^2)} \tag{14.16}$$

or, in our example,

$$r_w^2 = \frac{(45.6)^2}{(32.0)(68.8)} = .9445$$

If we multiply both the numerator and the denominator of the last term on the right in (14.10) by Σy_w^2, we see that the error sum of squares, or SS_2, will then be

$$SS_2 = \Sigma y_w^2 - r_w^2 \Sigma y_w^2 = \Sigma y_w^2(1 - r_w^2) \tag{14.17}$$

or

$$SS_2 = 68.8(1 - .9445) = 3.82$$

Because $\Sigma y_w^2 = SS_W$ or the error sum of squares in the analysis of variance, SS_2 and SS_W will be equal only if r_w is equal to zero. The larger the value of r_w, the smaller SS_2 will be in comparison with SS_W. It is because $r_w^2 = .9445$ is so large that SS_2 is so small compared with SS_W in this example. In actual experiments, values of r_w between .40 and .85 would be considerably more typical than the value in our hypothetical example.

Why is SS_5 (77.76) in the analysis of covariance so much larger than SS_T (57.6) in the analysis of variance? Rewriting (14.14) for SS_5, we see that

$$SS_5 = \Sigma y_t^2 - \left[\frac{(\Sigma xy_{tot})^2}{\Sigma x_{tot}^2} - \frac{(\Sigma xy_w)^2}{\Sigma x_w^2}\right] \tag{14.18}$$

Because Σy_t^2 is equal to SS_T or the treatment sum of squares in the analysis of variance, SS_5 can be equal to SS_T only if the two terms within brackets are equal. Substituting the appropriate values from Table 14.3 in (14.18), we have

$$SS_5 = 57.6 - \left[\frac{(38.4)^2}{32.9} - \frac{(45.6)^2}{32.0}\right] = 77.761$$

Because the product sum for treatments (Σxy_t) is negative, whereas the product sum within treatments (Σxy_w) is positive, we must have $\Sigma xy_w > \Sigma xy_{tot}$. Because it is also true that $\Sigma x_{tot}^2 \geq \Sigma x_w^2$, it follows that, in this example,

$$\frac{(\Sigma xy_{tot})^2}{\Sigma x_{tot}^2} < \frac{(\Sigma xy_w)^2}{\Sigma x_w^2}$$

and when this happens, SS_5 will be larger than SS_T.

If both the treatment product sum and the within treatment product sum are positive, then SS_5 can be smaller than SS_T. If the means on the covariate X for all treatment groups are exactly equal, then SS_5 will also be exactly equal to SS_T.

The chapter on the analysis of covariance in most textbooks is one of the most difficult chapters for students to understand, so if you have had difficulty with the material, do not despair. We will approach the same example from the viewpoint of multiple regression, and, hopefully, this will help to make more clear the nature of the analysis of covariance.

14.7 Multiple Regression Analysis of the Data in Table 14.2

In Table 14.4 we repeat the values of Y for our example with $k = 2$ treatments. We have let X_1 be the values of the covariate and X_2 be the vector for coding the two treatment groups. Vector $X_3 = X_1X_2$ is a product vector and corresponds to a vector that carries an interaction effect. In this instance, X_3 carries the interaction between the

TABLE 14.4 Values of Y and of the covariate (X_1) given in Table 14.2 with a vector for treatments (X_2) and the product vector $X_3 = X_1X_2$

X_1	X_2	X_3	Y
5	1	5	11
3	1	3	9
1	1	1	5
4	1	4	8
6	1	6	12
2	−1	−2	1
6	−1	−6	7
4	−1	−4	3
7	−1	−7	8
3	−1	−3	2

covariate and the treatments. We do a sequential multiple regression analysis with the X vectors entered in the order shown in Table 14.4.

With a sequential analysis we have

$$R^2_{Y.1} = .35458$$
$$R^2_{Y.12} = .96978$$
$$R^2_{Y.123} = .97096$$

14.8 Test for Interaction

In the analysis of covariance, SS_1 is a residual sum of squares based on the variation within each treatment group about the regression line fitted separately for each group. This residual sum of squares is the minimum possible value we could obtain, because we have minimized the residual sum of squares within each of the treatment groups. The counterpart of SS_1 in the multiple regression analysis is $1 - R^2_{Y.123}$. We observe, for example, that

$$(1 - R^2_{Y.123})SS_{tot} = (1 - .97096)(126.4) = 3.671$$

is equal to SS_1 in the analysis of covariance.*

We have $R^2_{Y.12} = .96978$, which represents the proportion of the total sum of squares that can be attributed to the covariate, X_1, *and* the treatments, X_2. It does not take into account the interaction vector X_3. Then $1 - R^2_{Y.12}$ is a residual sum of squares that does not take into account the variation attributable to the interaction vector X_3. It is the counterpart of SS_2 in the analysis of covariance. Note, for example, that

$$(1 - R^2_{Y.12})SS_{tot} = (1 - .96978)(126.4) = 3.820$$

is equal to SS_2 in the analysis of covariance.

In the analysis of covariance we had $SS_3 = SS_2 - SS_1$, and the counterpart of SS_3 in the multiple regression analysis is

$$(1 - R^2_{Y.12}) - (1 - R^2_{Y.123}) = R^2_{Y.123} - R^2_{Y.12}$$

and the difference is the squared semipartial correlation coefficient $r^2_{Y(3.12)}$ and represents the proportion of the total sum of squares that can be attributed to the interaction vector X_3 after taking into account

* Note, for example, that $R^2_{Y.123}$ takes into account the variation attributable to the covariate, X_1, the treatments, X_2, and the interaction, X_3, and that $1 - R^2_{Y.123}$ represents the proportion of the total sum of squares that cannot be accounted for by any of the X vectors.

the covariate and the treatments. Note, for example, that

$$(R^2_{Y.123} - R^2_{Y.12})SS_{tot} = (.97096 - .96978)(126.4) = .149$$

which is equal to SS_3 in the analysis of covariance.

For the test of significance of $r^2_{Y(3.12)} = R^2_{Y.123} - R^2_{Y.12}$ we have

$$F = \frac{R^2_{Y.123} - R^2_{Y.12}}{(1 - R^2_{Y.123})/(n - k - 1)} \tag{14.19}$$

where $k = 3$ is the number of X vectors. For our example we have

$$F = \frac{.00118}{(1 - .97096)/6} = .2438$$

which is equal, within rounding errors, to the F test of SS_3 in the analysis of covariance.

In the analysis of covariance we said that the F test of SS_3 was a test of significance of the difference between the two regression coefficients b_1 and b_2. In the multiple regression analysis the test of significance of $r^2_{Y(3.12)}$ is a test of the interaction of the treatments with the covariate. In what sense are the two tests comparable? Recall that in our earlier discussion of interaction in Chapter 9, we said that one way to examine a two-factor interaction is to plot the means or sums for each level of one factor against the levels of the other factor. If the lines are parallel, then the interaction sum of squares will be equal to zero. In the analysis of covariance we can think of the various values of the covariate X as representing the different levels of a factor. We then plot the individual values of Y for each treatment group against the values of X. We then fit the regression lines separately for each treatment group. If these two lines, one corresponding to the regression line for Treatment 1 and the other to the regression line for Treatment 2 in our example, are parallel, then there will be no interaction between the treatments and the covariate. But the only way in which the two regression lines can be parallel is if they have the same slope, that is, only if b_1 and b_2 are equal. Thus, the test of significance of SS_3 in the analysis of covariance is essentially a test to determine whether there is any interaction between the covariate and the treatments. To state that b_1 and b_2 differ significantly is to say that there is an interaction between the treatments and the covariate.

14.9 Test for Treatment Effects

In calculating SS_4 in the analysis of covariance, we took the regression of Y on X_1, the covariate, ignoring any treatment differences and any possible interaction. The counterpart of SS_4 in the multiple regression analysis is $1 - R^2_{Y.1}$, which represents the residual variation when only the covariate has been taken into account. It includes all of the variation attributable to the treatments and the possible interaction of treatments with the covariate. Then we have

$$(1 - R^2_{Y.1})SS_{tot} = (1 - .35458)(126.4) = 81.581$$

which is equal to SS_4 in the analysis of covariance.

In the analysis of covariance, SS_5 was defined as the difference between SS_4 and SS_2. In the multiple regression analysis this corresponds to

$$(1 - R^2_{Y.1}) - (1 - R^2_{Y.12}) = R^2_{Y.12} - R^2_{Y.1}$$

and $R^2_{Y.12} - R^2_{Y.1}$ is the squared semipartial correlation coefficient $r^2_{Y(2.1)}$ and represents the proportion of the total sum of squares attributable to X_2, the treatment vector, after we have taken into account the covariate, X_1. In our example we have

$$r^2_{Y(2.1)} = R^2_{Y.12} - R^2_{Y.1} = .96978 - .35458 = .61520$$

and we note that

$$(R^2_{Y.12} - R^2_{Y.1})SS_{tot} = (.61520)(126.4) = 77.761$$

which is equal to SS_5 in the analysis of covariance.

For the test of significance of $r^2_{Y(2.1)}$ we have

$$F = \frac{R^2_{Y.12} - R^2_{Y.1}}{(1 - R^2_{Y.12})/(n - k - 1)} \qquad (14.20)$$

where $k = 2$ is the number of X vectors on which $R^2_{Y.12}$ is based. In our example we have

$$F = \frac{.61520}{(1 - .96978)/7} = 142.502$$

which is equal, within rounding errors, to the F test of SS_5 in the analysis of covariance.

14.10 Multiple Covariates

It is not necessary, as we pointed out at the beginning of the chapter, that the covariate be the same measure as the dependent variable. It can be any variable that we have reason to believe will be correlated with the dependent variable. Nor are we limited to a single covariate. We might, for example, have two or three covariates in anticipation that the multiple correlation of the dependent variable with the set of covariates will be substantially greater than the correlation of the dependent variable with any one of the covariates. The multiple regression analysis described in the chapter can easily be applied in the case of multiple covariates.

Suppose, for example, that we have two covariates, X_1 and X_2, and three treatments coded by X_3 and X_4. Then we will have the product vectors $X_5 = X_1X_3$, $X_6 = X_1X_4$, $X_7 = X_2X_3$, and $X_8 = X_2X_4$. For the test of significance of the interaction we will have

$$F = \frac{(R^2_{Y.12345678} - R^2_{Y.1234})/4}{(1 - R^2_{Y.12345678})/(n - 8 - 1)}$$

If the interaction is not significant, then we will test the treatment effect for significance with

$$F = \frac{(R^2_{Y.1234} - R^2_{Y.12})/2}{(1 - R^2_{Y.1234})/(n - 4 - 1)}$$

To generalize these tests for any number of covariates and any number of treatments, we let A = the number of covariates, B = the number of treatment vectors, and C = the number of product vectors. Then the test of the interaction effect will be

$$F = \frac{(R^2_{Y.ABC} - R^2_{Y.AB})/C}{(1 - R^2_{Y.ABC})/(n - k - 1)} \tag{14.21}$$

where $k = A + B + AB$, or the total number of X vectors, and the test of significance of the treatment effect will be

$$F = \frac{(R^2_{Y.AB} - R^2_{Y.A})/B}{(1 - R^2_{Y.AB})/(n - k - 1)} \tag{14.22}$$

where $k = A + B$.

Exercises

14.1 The following questions refer to the relationships between the analysis of covariance and a multiple regression analysis of the same data when we have a single covariate and two treatments.

(a) If SS_1 is divided by SS_{tot}, what will this be equal to in a multiple regression analysis?

(b) If SS_2 is divided by SS_{tot}, what will this be equal to in a multiple regression analysis?

(c) If SS_3 is divided by SS_{tot} what will this be equal to in a multiple regression analysis?

(d) If SS_4 is divided by SS_{tot}, what will this be equal to in a multiple regression analysis?

(e) If SS_5 is divided by SS_{tot}, what will this be equal to in a multiple regression analysis?

14.2 Will SS_1 always be smaller than SS_2? Explain why or why not.

14.3 Under what conditions will SS_5 be equal to the treatment sum of squares in the analysis of variance of the Y variable?

14.4 Under what conditions will SS_2 be equal to the within treatment sum of squares in the analysis of variance of the Y variable?

14.5 In Table 14.2 we gave the values of X and Y for two treatments. We now add a third treatment. The values of X and Y for this treatment are as follows:

Treatment 3	
X	*Y*
1	10
4	13
5	16
3	12
6	17

Now complete the analysis of covariance for the three treatments. Note that you will not have to repeat the calculations for the two treatments in Table 14.2. They are already available.

(a) What is the value of SS_1 for the three treatments?
(b) What is the value of SS_2 for the three treatments?
(c) Calculate the values of b_1, b_2, b_3. Do they differ significantly?
(d) What is the value of SS_4?
(e) Are there significant treatment effects?

14.6 With the three treatment groups as described in Exercise 14.5, do a multiple regression analysis of the data. Let X_1 be the covariate. Code X_2 with 1 for all observations in Treatment 1 and zero otherwise. Code X_3 with 1 for all observations in Treatment 2 and zero otherwise. Let $X_4 = X_1X_2$ and $X_5 = X_1X_3$. Do a sequential analysis with the X vectors entered in the regression equation in the order given. The values of $R^2_{YY'}$ will be: $R^2_{Y.1} = .18122$, $R^2_{Y.12} = .18315$, $R^2_{Y.123} = .98294$, $R^2_{Y.1234} = .98342$, and $R^2_{Y.12345} = .98344$.

(a) Test to determine if there is a significant interaction between the treatments and the covariate. Is this F ratio equal to the value obtained in Exercise 14.5 for the test of significance of differences in the three regression coefficients?

(b) Test to determine if there are significant treatment effects. Is this F ratio equal to the one obtained in Exercise 14.5 for the test of significance for treatment effects?

14.7 Explain each of the following concepts:

covariate

homogeneity of regression coefficients

total product sum

treatment product sum

within treatment product sum

15

Conditional Fixed-Effect Models

15.1 A Conditional Fixed-Effect Model

We have throughout the book emphasized that we were concerned with a fixed-effect model in which the X vectors and variables were fixed constants. This is a reasonable model for experimental data in which the treatments and levels of treatment factors seldom represent a random selection from some larger population. With a fixed-effect model, only the values of Y are subject to random variation.

With nonexperimental data, the values of X variables are often not fixed. For example, if X_1 represents high-school grades, X_2 represents scores on an aptitude test, and Y represents first-year college grades, X_1 and X_2 are subject to random variation as well as Y. Similarly, in almost all analysis of covariance designs, the covariate X_1, whether it be a pretest on the dependent variable or some individual difference variable, will also be subject to random variation.

In those examples cited in the chapters on multiple regression and the analysis of covariance, where the values of X may be considered random, we have assumed a *provisional* or *conditional fixed-effect model.* Given that we have a random sample of X values, we then regarded the values of X as fixed and the values of Y as conditional on the values of X observed. The validity of any conclusions drawn from the analysis of the data strictly apply only to the actual X values investigated. But if the study were to be repeated, the values of X in the repetition should not differ drastically from those in the original study if both are random samples from the same population. That is, we might expect the range of the X values in successive repetitions of the study to be fairly comparable. Thus, it does not seem unreasonable to believe that the conclusions based on a random sample of X values, which are then regarded as fixed, will hold for additional random samples of X values with approximately the same range.

With a conditional fixed-effect model it will almost always be true that product vectors involving the X variables will not be orthogonal to the vectors making up the product. If the product vector $X_3 = X_1X_2$ is not orthogonal to X_1 and X_2, then it is ordinarily not the correlation of X_3 with Y that is of interest but rather the correlation of X_3 with Y after we have partialed X_1 and X_2 from X_3 or, in other words, the semipartial correlation $r_{Y(3.12)}$.

15.2 Sequential Ordering of X Vectors and Variables

In all of the examples cited in the book we used a sequential ordering of the X variables or vectors in a multiple regression analysis. With an orthogonal design and with all X vectors mutually orthogonal, the sequential ordering is primarily a matter of convenience. For example, in a two-factor experiment we wanted the set of X vectors corresponding to the levels of A to be together, the set of X vectors corresponding to the levels of B to be together, and the set of X vectors corresponding to the $A \times B$ interaction to be together.

If X vectors are not orthogonal, then a sequential ordering of the X vectors makes it easier to find certain semipartial correlations of interest. For example, if X_1 is a quantitative variable, we may be interested not only in the linear relationship between X_1 and Y but also in the product vectors $X_2 = X_1X_1 = X_1^2$ and $X_3 = X_1X_2 = X_1^3$, which carry the quadratic and cubic components. Now X_1, X_2, and X_3 will, in general, be highly correlated, and X_2 and X_3 may correlate almost as highly with Y as X_1. It is not, however, the correlations of X_2 and X_3 with Y that are of interest but rather the semipartial correlations $r_{Y(2.1)}$ and $r_{Y(3.12)}$. For example, $r^2_{Y(2.1)} = R^2_{Y.12} - R^2_{Y.1}$ is the proportion of the total sum of squares attributable to the quadratic component, X_2, after we have already taken into account the linear component, X_1. Similarly, $r^2_{Y(3.12)} = R^2_{Y.123} - R^2_{Y.12}$ is the proportion of the total sum of squares attributable to the cubic component, X_3, after we have already taken into account the linear and quadratic components, X_1 and X_2.

15.3 Orthogonal Product Vectors

If a product vector $X_3 = X_1X_2$ is orthogonal to X_1 and X_2, then $r_{Y3} = r_{Y(3.12)}$, and we need not be concerned with partialing X_1 and X_2 from X_3. For example, consider the 2 × 2 factorial experiment in Table 15.1. This is an orthogonal design, because we have $n = 2$ observations for each of the treatment combinations. For the data in Table 15.1 it is easy to show that

$$SS_T = SS_A + SS_B + SS_{AB}$$

or

$$12 = 8 + 2 + 2$$

and $SS_{tot} = 14$. Then

$$R^2_{Y.123} = \tfrac{12}{14} = \tfrac{8}{14} + \tfrac{2}{14} + \tfrac{2}{14} = .857143$$

We also have

$$SS_A = SS_{tot}r^2_{Y1} = 14\frac{(14-6)^2}{(8)(14)} = 8$$

TABLE 15.1 Orthogonal coding for a 2 × 2 factorial experiment

	Subjects	X_1	X_2	X_3	Y
A_1B_1	1	1	1	1	2
A_1B_1	2	1	1	1	3
A_1B_2	3	1	−1	−1	4
A_1B_2	4	1	−1	−1	5
A_2B_1	5	−1	1	−1	1
A_2B_1	6	−1	1	−1	2
A_2B_2	7	−1	−1	1	1
A_2B_2	8	−1	−1	1	2

and

$$SS_B = SS_{tot} r_{Y2}^2 = 14 \frac{(8 - 12)^2}{(8)(14)} = 2$$

Now note that $X_3 = X_1 X_2$ is a product vector but that X_3 is also orthogonal to X_1 and X_2, and, because it is, we have $r_{Y3}^2 = r_{Y(3.12)}^2$. Thus, with X_3 orthogonal to X_1 and X_2 we have

$$SS_{AB} = SS_{tot} r_{Y3}^2 = 14 \frac{(8 - 12)^2}{(8)(14)} = 2$$

15.4 Nonorthogonal Product Vectors

We now examine the same experiment with a new set of X vectors, as shown in Table 15.2. In this instance X_1 and X_2 are vectors with dummy coding for the two levels of A and the two levels of B, respectively, and $X_3 = X_1 X_2$ is a product vector that *carries* the $A \times B$ interaction.* X_3 is not itself the $A \times B$ interaction, because, as we shall see, X_3 is not orthogonal to X_1 and X_2.

We observe, however, that X_1 and X_2 are arthogonal because

$$\Sigma(X_1 - \bar{X}_1)(X_2 - \bar{X}_2) = 2 - \frac{(4)(4)}{8} = 0$$

We also have

$$\Sigma(X_1 - \bar{X}_1)^2 = 4 - \frac{(4)^2}{8} = 2$$

and

$$\Sigma(X_1 - \bar{X}_1)(Y - \bar{Y}) = 14 - \frac{(4)(20)}{8} = 4$$

and

$$r_{Y1}^2 = \frac{(4)^2}{(2)(14)} = .57143$$

so that

$$SS_A = SS_{tot} r_{Y1}^2 = 14 \frac{(4)^2}{(2)(14)} = 8$$

as before.

* For a detailed discussion of the role of product vectors in multiple regression, see Cohen (1978).

TABLE 15.2 Dummy coding for the levels of A (X_1) and B (X_2) with $X_3 = X_1X_2$

	Subjects	X_1	X_2	X_3	Y
A_1B_1	1	1	1	1	2
A_1B_1	2	1	1	1	3
A_1B_2	3	1	0	0	4
A_1B_2	4	1	0	0	5
A_2B_1	5	0	1	0	1
A_2B_1	6	0	1	0	2
A_2B_2	7	0	0	0	1
A_2B_2	8	0	0	0	2

Similarly, for X_2 we have

$$\Sigma x_2^2 = \Sigma(X_2 - \bar{X}_2)^2 = 2$$

and

$$\Sigma(X_2 - \bar{X}_2)(Y - \bar{Y}) = 8 - \frac{(4)(20)}{8} = -2$$

and

$$r_{Y2}^2 = \frac{(-2)^2}{(2)(14)} = .14286$$

so that

$$SS_B = SS_{tot}r_{Y2}^2 = 14\frac{(-2)^2}{(2)(14)} = 2$$

as before.

We see, however, that $X_3 = X_1X_2$ is *not* orthogonal to X_1 and X_2, because

$$\Sigma(X_1 - \bar{X}_1)(X_3 - \bar{X}_3) = 2 - \frac{(4)(2)}{8} = 1$$

and

$$\Sigma(X_2 - \bar{X}_2)(X_3 - \bar{X}_3) = 2 - \frac{(4)(2)}{8} = 1$$

Because $X_3 = X_1X_2$ is not orthogonal to X_1 and X_2, X_3 is not the $A \times B$ interaction. To obtain the $A \times B$ interaction we must partial

X_1 and X_2 from X_3; that is, we want $r_{Y(3.12)}$, not r_{Y3}. In fact, we see that with dummy coding for A and B in the example under discussion, we have

$$\Sigma(X_3 - \bar{X}_3)(Y - \bar{Y}) = 5 - \frac{(2)(20)}{8} = 0$$

and, consequently, $r_{Y3} = 0$. Thus,

$$R^2_{Y.123} \neq r^2_{Y1} + r^2_{Y2} + r^2_{Y3}$$

but instead

$$R^2_{Y.123} = r^2_{Y1} + r^2_{Y2} + r^2_{Y(3.12)}$$

We know that

$$R^2_{Y.123} = \frac{SS_T}{SS_{tot}} = \frac{12}{14} = .85714$$

and because X_1 and X_2 are orthogonal we also know that

$$R^2_{Y.12} = r^2_{Y1} + r^2_{Y2} = .57143 + .14286 = .71429$$

Then we have

$$r^2_{Y(3.12)} = R^2_{Y.123} - R^2_{Y.12} = .85714 - .71429 = .14285$$

and it will be true that

$$\begin{aligned} SS_{tot}R^2_{Y.123} &= SS_{tot}(r^2_{Y1} + r^2_{Y2} + r^2_{Y(3.12)}) \\ &= SS_A + SS_B + SS_{AB} \end{aligned}$$

In general, the correlations of product vectors with Y are of no intrinsic interest except in those cases where the product vectors are orthogonal to the vectors making up the product.

15.5 Trait–Treatment Interactions

In Chapter 14, on the analysis of covariance, our primary interest was in the difference between the treatments. The covariate X was included in the experiment because we anticipated that it would result in a reduction of the error sum of squares and we would have a more sensitive test of the treatment effect. We definitely were not interested in finding a significant interaction between the covariate and the treatments, because the validity of the F test for the treatment effect assumes that the regression coefficients for the various treatment groups do not differ significantly.

There are, however, experiments in which the interaction of a covariate with treatments is of primary interest. For example, a not uncommon design in social psychology involves dividing subjects into two or three groups on the basis of scores on a personality scale: above average and below average or above average, average, and below average. In educational psychology the subjects may be divided into above average and below average or above average, average, and below average groups on the basis of scores on an aptitude test. We shall refer to these individual difference variables as traits.

Within each level of a trait, subjects are assigned at random to two or more treatment groups. Primary interest is not in the treatment differences but rather in the possible interaction between the trait and the treatments. Let us assume that an experiment of this kind consists of three levels of a trait (A) and two treatments (B) with $n = 10$ subjects for each trait–treatment combination (A_iB_j). The analysis of variance for this experiment will be as summarized below.

Source of Variation	d.f.
Trait (A)	$a - 1$
Treatments (B)	$b - 1$
A × B	$(a - 1)(b - 1)$
Within	$ab(n - 1)$

Figure 15.1 shows some of the possible outcomes of the experiment with respect to the trait–treatment interaction. In Figure 15.1(a) the interaction is ordinal but linear, and in Figure 15.1(b) the interaction is disordinal but linear. Figure 15.1(c) indicates that the interaction includes a quadratic component. If the values of A_1, A_2, and A_3 represented equally spaced values of trait differences, we could use coefficients for orthogonal polynomials to partition the interaction sum of squares into the two orthogonal components, the linear and quadratic.

Let us consider the same experiment in terms of a multiple regression analysis. Suppose that the subjects are assigned completely at random to the two treatments. If the number of subjects is the same as in the analysis of variance experiment described above, we would have thirty subjects for each treatment. For each subject we would also have a measure on the trait of interest. Note that these trait measures are considerably more precise than when they are reduced to three classes: above average, average, and below average, as used in the analysis of variance. The trait measures serve as a covariate X_1. The two treatments can be coded with a second vector X_2, and the interaction can be carried by the product vector $X_3 = X_1X_2$. Then to determine whether b_1 and b_2

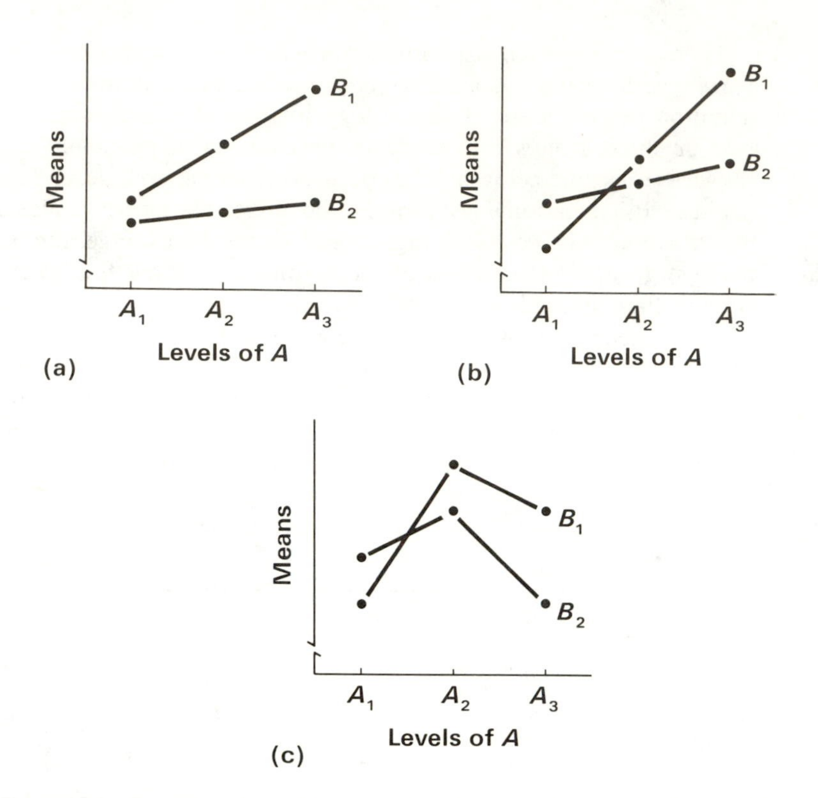

Figure 15.1 Examples of possible interactions between three levels of a trait (A) and two treatments (B).

differ significantly, or whether there is a significant linear interaction between the trait and the treatments, we would have

$$F = \frac{R^2_{Y.123} - R^2_{Y.12}}{(1 - R^2_{Y.123})/(n - k - 1)} = \frac{r^2_{Y(3.12)}}{(1 - R^2_{Y.123})/(n - k - 1)} \qquad (15.1)$$

where $k = 3$ is the number of X vectors.

In the analysis of covariance, the F test of (15.1) is a planned "preliminary" test to determine whether we should go on to the test of the treatment effect

$$F = \frac{R^2_{Y.12} - r^2_{Y1}}{(1 - R^2_{Y.12})/(n - k - 1)} = \frac{r^2_{Y(2.1)}}{(1 - R^2_{Y.12})/(n - k - 1)} \qquad (15.2)$$

where $k = 2$ is the number of X vectors after eliminating X_3. In other words, assuming that the F test of (15.1) was not significant, we dropped the interaction term from our model.

As we stated before, the test of a treatment effect in the presence of a significant interaction may or may not be meaningful.* In the analysis of covariance it is obvious that if the slopes of the regression lines for the treatments differ significantly, then the magnitude of the difference between the treatment means will depend on the value of the covariate. For example, with high values of the covariate the difference between the treatment means may be quite large, whereas for low values of the covariate the difference may be quite small, assuming that the regression lines do not cross within the range of the values of the covariate studied. If the regression lines cross, then the direction of the difference between the treatment means will be reversed for high and low values of the covariate.

If we wished to consider the possibility that there may be a quadratic component of the trait–treatment interaction, we would include the following X vectors:

$$X_1 = \text{the covariate: } A$$
$$X_2 = \text{treatments: } B$$
$$X_3 = X_1X_2 = \text{linear component of } A \times B$$
$$X_4 = X_1X_1 = \text{quadratic component of } A$$
$$X_5 = X_2X_4 = \text{quadratic component of } A \times B$$

Because the linear and quadratic terms will be correlated, we would assign priority to the more simple, or linear, interaction. Then to test the quadratic component of the $A \times B$ interaction for significance, we would calculate

$$F = \frac{R^2_{Y.12345} - R^2_{Y.1234}}{(1 - R^2_{Y.12345})/(n-k-1)} = \frac{r^2_{Y(5.1234)}}{(1 - R^2_{Y.12345})/(n-k-1)} \tag{15.3}$$

where $k = 5$ is the number of X vectors. If the quadratic component of the $A \times B$ interaction were not significant, then we would drop X_5 and test for the quadratic component of A by means of

$$F = \frac{R^2_{Y.1234} - R^2_{Y.123}}{(1 - R^2_{Y.1234})/(n-k-1)} \tag{15.4}$$

where $k = 4$ is the number of X vectors.

If the quadratic component of A were not significant, then we would have our simple analysis of covariance model with three X vectors. We

* See the discussion in Section 9.3.

would then use (15.1) to test for the linear component of $A \times B$. If the linear component of $A \times B$ were not significant, we would have the test for the treatment effects as given by (15.2).

15.6 Pseudo-Orthogonal Designs

Suppose an investigator is interested in the influence of two traits or individual difference variables, A and B, on a dependent variable Y. Each of the trait variables is dichotomized at the median of the distribution of scores to obtain high and low groups, A_1 and A_2 and B_1 and B_2. Then this design will resemble a 2×2 factorial experiment with two factors, A and B, each with two levels, as shown in Table 15.3.

If the two traits are positively correlated, then we will expect to have for any random sample more subjects in cells A_1B_1 and A_2B_2 than in A_2B_1 and A_1B_2. Similarly, if the two traits are negatively correlated, we will expect to have more subjects in cells A_1B_2 and A_2B_1 than in cells A_1B_1 and A_2B_2. Because scores on almost all aptitude tests are positively correlated and scores on personality scales are either positively or negatively correlated, then in order to have a balanced or orthogonal design with an equal number of subjects in each cell, some subjects will have to be discarded randomly from the cells with large frequencies. Assuming the investigator obtains an equal number of subjects for each cell, the data are then analyzed as if they had been obtained from a 2×2 factorial experiment. The results of the analysis of variance can, however, be quite misleading, as Humphreys and Fleishman (1974) have shown. For one thing, the data will be analyzed as if the two traits are uncorrelated when, in fact, they are not. Humphreys and Fleishman have described designs of this kind as "pseudo-orthogonal" designs.

The investigation of two individual differences variables, A and B, and a dependent variable Y can be done more appropriately with a

TABLE 15.3 Combinations of high and low levels of two traits obtained by dichotomizing the distribution of scores for each trait at the median

	Low: B_2	High: B_1
High: A_1	A_1B_2	A_1B_1
Low: A_2	A_2B_2	A_2B_1

multiple regression analysis. We let X_1 be the vector of scores measuring trait A and X_2 be the vector of scores measuring trait B. Note, first of all, that no subjects need to be discarded, as was necessary with the pseudo-orthogonal design. Second, the measures of the traits are more precise than the categorizations of these with two groups, high and low. For the interaction of A and B we have the product vector $X_3 = X_1X_2$, or simply the product of the scores on the two scales or tests.

We do a sequential analysis with the X vectors entered into the regression equation in order: X_1, X_2, and X_3. Now X_3 will not be orthogonal to X_1 and X_2 and will not itself be the $A \times B$ interaction. It simply carries the $A \times B$ interaction. For the $A \times B$ interaction we want the semipartial correlation $r_{Y(3.12)}$. Thus, the test of significance of the $A \times B$ interaction will be given by

$$F = \frac{R^2_{Y.123} - R^2_{Y.12}}{(1 - R^2_{Y.123})/(n - k - 1)} = \frac{r^2_{Y(3.12)}}{(1 - R^2_{Y.123})/(n - k - 1)} \tag{15.5}$$

If Model 8, described in Chapter 12, is the relevant population model, then the tests of significance of A and B will be given by*

$$F = \frac{R^2_{Y.123} - R^2_{Y.23}}{(1 - R^2_{Y.123})/(n - k - 1)} = \frac{r^2_{Y(1.23)}}{(1 - R^2_{Y.123})/(n - k - 1)} \tag{15.6}$$

and

$$F = \frac{R^2_{Y.123} - R^2_{Y.13}}{(1 - R^2_{Y.123})/(n - k - 1)} = \frac{r^2_{Y(2.13)}}{(1 - R^2_{Y.123})/(n - k - 1)} \tag{15.7}$$

respectively.

The F test defined by (15.5) is a test to determine whether X_1 contributes significantly to the regression sum of squares over and above that already accounted for by X_2 and X_3. Similarly, the F test defined by (14.6) is a test to determine whether X_2 contributes significantly to the regression sum of squares over and above that already accounted for by X_1 and X_3.

15.7 The Analysis of Covariance for a Pretest–Posttest Design

An experimental design frequently used in the behavioral sciences is the pretest–posttest design. In this design, subjects are divided randomly into two groups. Both groups of subjects are given a pretest; that is, a

* Other models are possible for testing the significance of A and B. See the discussion in Chapter 12 and the references cited.

measure is obtained for each subject on the dependent variable *prior* to the administration of the treatments. After the pretest, one group of subjects receives treatment A_1 and the other treatment A_2. Both groups of subjects are then given a posttest on the dependent variable.

We let B_1 correspond to the pretest, or the measures obtained prior to the administration of the treatments, and B_2 to the posttest, or the measures obtained after the administration of the treatments. This design may *appear* to resemble a split-plot design with subjects randomly assigned to A_1 and A_2 and with two repeated measures on each subject, B_1 and B_2. In a split-plot design, however, the measures on both B_1 and B_2 are obtained in the presence of the treatments A_1 and A_2, and A_1 and A_2 can have an influence on both B_1 and B_2. In the pretest–posttest design, on the other hand, it is obvious that A_1 and A_2 cannot have any influence on B_1, because the B_1 measures were obtained *prior* to the administration of A_1 and A_2.

Table 15.4(a) provides a notation for the four cell means and also for the four marginal means. If the subjects have been randomly assigned to A_1 and A_2, and if both groups of subjects are given a pretest in the absence of any treatments, then we would expect the mean for A_1B_1 and the mean for A_2B_1 to be comparable; for simplicity we have let them be equal. Table 15.4(b) indicates that the treatments have had an effect; the mean for A_1B_2 is 40.0 and the mean for A_2B_2 is 30.0, a difference of 10.0 between the two means.

We would like to test for the treatment effect and it may seem as if the appropriate test is $F = MS_A/MS_{S(A)}$, but it is not. This is a test of the difference between the two means 30.0 and 25.0, a difference of only 5.0 between the two means. The inclusion of the pretest means masks or diminishes the effect of the two treatments. The test we should make, if we wish to determine whether there is a significant difference between A_1 and A_2 with the split-plot design, is

$$F = \frac{MS_{AB}}{MS_{S(A)B}}$$

TABLE 15.4 Cell means and marginal means for a pretest–posttest design

(a)

	B_1	B_2	Means
A_1	$\bar{Y}_{11}$	$\bar{Y}_{12}$	$\bar{Y}_{1.}$
A_2	$\bar{Y}_{21}$	$\bar{Y}_{22}$	$Y_{2.}$
Means	$\bar{Y}_{.1}$	$\bar{Y}_{.2}$	$\bar{Y}$

(b)

	B_1	B_2	Means
A_1	20.0	40.0	30.0
A_2	20.0	30.0	25.0
Means	20.0	35.0	27.5

The F test of the $A \times B$ interaction can be shown to be equivalent to a test of the null hypothesis

$$(\mu_{12} - \mu_{11}) - (\mu_{22} - \mu_{21}) = 0$$

The sample mean estimates of these population means are, as shown in Table 15.4(b)

$$(40.0 - 20.0) - (30.0 - 20.0) = 10.0$$

and it is this difference that we want to test for significance.

Although the pretest–posttest design can be analyzed as if it were a split-plot design, we believe that a more appropriate analysis is the analysis of covariance, or its multiple regression counterpart, in which the pretest (B_1) is treated as a covariate X_1.*

Exercises

15.1 Suppose that we have a factor A with $a = 3$ levels and another factor B with $b = 2$ levels with an equal number of subjects for each A_iB_j treatment combination. We use dummy (1, 0) coding for the levels of A and the levels of B.

(a) Will X_1 and X_2 used to code A be orthogonal?

(b) Will X_3 used to code B be orthogonal to X_1 and X_2?

(c) Will the product vectors $X_4 = X_1X_3$ and $X_5 = X_2X_3$ be orthogonal to X_1, X_2, and X_3?

(d) If the vectors are entered sequentially in the regression equation, show how you will obtain SS_A, SS_B, and SS_W.

15.2 For the experiment described in Exercise 15.1, we use effect coding $(1, 0, -1)$ and $(0, 1, -1)$ for the three levels of A and $(1, -1)$ for the two levels of B.

(a) Will X_1 and X_2 be orthogonal?

(b) Will X_3 be orthogonal to X_1 and X_2?

(c) Will the product vectors $X_4 = X_1X_3$ and $X_5 = X_2X_3$ be orthogonal to X_1, X_2, and X_3?

(d) If the vectors are entered sequentially in the regression equation, show how you will obtain SS_A, SS_B, and SS_W.

* For a more complete discussion of the analysis of the pretest–posttest design, see Huck and McLean (1975), Levin and Marascuilo (1977), and Edwards (1984b).

15.3 We have a $2 \times 2 \times 2$ factorial experiment with an equal number of observations for each $A_iB_jC_k$ treatment combination. We code the two levels of each factor $(1, -1)$.

(a) Will X_1, X_2, and X_3 be mutually orthogonal?

(b) Will the product vectors $X_4 = X_1X_2$, $X_5 = X_1X_3$, $X_6 = X_2X_3$, $X_7 = X_1X_2X_3$ be orthogonal to each other?

(c) Are all X vectors mutually orthogonal?

(d) What sum of squares will be given by $SS_{tot}r^2_{Y7}$?

15.4 We have an experiment with one covariate and three treatments.

(a) Show how you would test the null hypothesis $\beta_1 = \beta_2 = \beta_3$ for significance with a multiple regression analysis.

(b) If the regression coefficients differ significantly, is there anything wrong with testing for the significance of the treatment effect? Explain why or why not.

(c) Suppose the treatment effect is signficant. What does this mean?

15.5 An experimenter assigns subjects completely at random to two treatments. The test of significance of the treatment mean square is significant in the analysis of variance. The experimenter happens to have scores on an aptitude test available for each subject and reanalyzes the data using the aptitude test score as a covariate. The interaction between the aptitude test scores and the treatments is significant.

(a) What should the experimenter do? After all, the results of the original study could have been submitted for publication. Should the reanalysis with the covariate be ignored?

(b) If the treatment effect is significant in the original analysis, what is the likely nature of the interaction in the analysis using the covariate?

15.6 Many experiments in the behavioral sciences are of the completely randomized kind with no covariate. The test of significance of the treatment effect in designs of this kind is $F = MS_T/MS_W$.

(a) If an experimenter fails to find a significant value of F, could this be the result of a significant interaction between some individual difference variable and the treatments? Explain why or why not.

(b) If it is the result of a trait–treatment interaction, what is the likely nature of the interaction? Explain why.

(c) If $F = MS_T/MS_W$ is not significant, and if the F test for the treatment effect is significant when the same data are reanalyzed using a covariate, what does this indicate? Explain why.

15.7 A pretest–posttest design is analyzed as if it were a split-plot design with two repeated measures. Explain why the test of significance of the treatment effect does not really reflect the magnitude of the difference between the two treatment means.

15.8 Explain each of the following concepts:

conditional fixed-effect model

linear component of interaction

nonorthogonal product vectors

orthogonal product vectors

pretest–posttest design

product vector

pseudo-orthogonal design

quadratic component of interaction

trait–treatment interaction

References

Anscombe, F. J. Graphs in statistical analysis. *The American Statistician*, 1973, 27, 17–21.

Appelbaum, M. I., and E. M. Cramer. Some problems in the nonorthogonal analysis of variance. *Psychological Bulletin*, 1974, 81, 335–343.

Carlson, J. E., and N. H. Timm. Analysis of nonorthogonal fixed-effects designs. *Psychological Bulletin*, 1974, 81, 563–570.

Cohen, J. Partialed products *are* interactions: Partialed powers *are* curve components. *Psychological Bulletin*, 1978, 85, 858–866.

Darlington, R. B. Reduced-variance regression. *Psychological Bulletin*, 1978, 85, 1238–1255.

Dixon, W. J., and M. B. Brown. *BMDP-77 Biomedical Computer Programs, P-Series.* Berkeley: University of California Press, 1977.

Draper, N. R., and H. Smith. *Applied Regression Analysis.* New York: Wiley, 1966.

Edwards, A. L. *An Introduction to Linear Regression and Correlation* (2nd ed.). New York: W. H. Freeman and Company, 1984a.

———. *Experimental Design in Psychological Research* (5th ed.). New York: Harper & Row, 1984b.

Gaito, J. Unequal intervals and unequal *n*'s in trend analysis. *Psychological Bulletin*, 1965, 63, 125–127.

Geisser, S., and S. W. Greenhouse. An extension of Box's results on the use of the *F* distribution in multivariate analysis. *Annals of Mathematical Statistics*, 1958, 29, 885–891.

Greenhouse, S. W., and S. Geisser. On methods in the analysis of profile data. *Psychometrika*, 1959, 24, 95–112.

Herr, D. G., and J. Gaebelein. Nonorthogonal two-way analysis of variance. *Psychological Bulletin*, 1978, 85, 207–216.

Horst, P. *Matrix Algebra for Social Scientists.* New York: Holt, Rinehart and Winston, 1963.

Horst, P., and A. L. Edwards. A mathematical proof of equivalence of an unweighted means analysis of variance and a Method 1 regression analysis for the 2^k unbalanced factorial experiment. *JSAS Catalog of Selected Documents in Psychology*, 1981, 3, 76 (Ms. No. 2354).

———. Analysis of nonorthogonal designs: The 2^k factorial experiment. *Psychological Bulletin*, 1982, 91, 190–192.

Huck, S. W., and R. A. McLean. Using a repeated measures ANOVA to analyze the data from a pretest–posttest design: A potentially confusing problem. *Psychological Bulletin*, 1975, 82, 511–518.

Humphreys, L. G., and A. Fleishman. Pseudo-orthogonal and other analysis of variance designs involving individual difference variables. *Journal of Educational Psychology*, 1974, 66, 464–472.

Huynh, H., and L. S. Feldt. Conditions under which mean square ratios in repeated measurements designs have exact *F*-distributions. *Journal of the American Statistical Association*, 1970, 65, 1582–1589.

Joe, G. W. Comment on Overall and Spiegel's "Least squares analysis of experimental data." *Psychological Bulletin*, 1971, 75, 464–466.

Levin, J. R., and L. A. Marascuilo. Post hoc analysis of repeated measures interactions and gain scores: Whither the inconsistency? *Psychological Bulletin*, 1977, 84, 247–248.

Lewis, C., and G. Keren. You can't have your cake and eat it too: Some considerations of the error term. *Psychological Bulletin*, 1977, 84, 1150–1154.

Nie, N. H., C. H. Hull, J. G. Jenkins, K. Steinbrenner, and D. H. Bent. *SPSS: Statistical Package Program for the Social Sciences* (2nd ed.). New York: McGraw-Hill, 1975.

Overall, J. E., and D. K. Spiegel. Concerning least squares analysis of experimental data. *Psychological Bulletin*, 1969, 72, 311–322.

———. Comments on Rawlings' nonorthogonal analysis of variance. *Psychological Bulletin*, 1973, 79, 164–167.

Pedhazur, E. J. Coding subjects in repeated measures designs. *Psychological Bulletin*, 1977, 34, 298–305.

Price, B. Ridge regression: Application to nonexperimental data. *Psychological Bulletin*, 1977, 84, 759–766.

Rawlings, R. R., Jr. Comments on the Overall and Spiegel paper. *Psychological Bulletin*, 1973, 79, 168–169.

Rouanet, H., and D. Lépine. Comparisons between treatments in a repeated-measurement design: ANOVA and multivariate methods. *The British Journal of Mathematical and Statistical Psychology*, 1970, 23, 147–163.

Searle, S. R. *Matrix Algebra Useful for Statistics*. New York: Wiley, 1982.

Speed, F. M., and C. J. Monlezun. Exact *F* tests for the method of unweighted means in a 2^k experiment. *The American Statistician*, 1979, 33, 15–18.

Answers to Selected Exercises

Answers to exercises involving calculations were, for the most part, obtained with a hand calculator using the full capacity of eight digits. Rounding errors are inevitable when calculations exceed the capacity of the calculator. Final answers have also been rounded. You should keep this in mind in comparing your answers with those reported here.

Chapter 1

1.1 (a) $a = 1$, $b = .5$
(b) $a = 3.25$, $b = -.25$
(c) $a = 4$, $b = -1.0$
(d) $a = -10$, $b = 5$

1.2 (a) 3, 5, 7
(b) 2.5, 2.0, 1.5
(c) 1.5, 2.5, 3.5
(d) −.8, −1.6, −2.4

1.3 7, 9, 11

1.4 $Y = X$

1.5 $Y = 10 + (-.5)X$

1.6 $Y = -5 + (-.4)X$

Chapter 2

2.1 (a) $r = .08$
(b) $r = 1.0$

2.2 (a) $r = .88$
(b) $r = 0$

2.3 $SS_{reg} = SS_{tot}$; $SS_{res} = 0$

2.4 If $s_X = s_Y$.

2.5 See text, Section 2.5.

2.6 $\Sigma(X - \bar{X})(Y - Y') = \Sigma x(y - bx)$
$= \Sigma xy - b\Sigma x^2 = 0$

2.7 $r = .707$

2.8 $r = .70$
2.9 $r = .50$
2.10 (a) $r = 0$
(b) $Y = X^2$
2.12 (a) $Y' = 1.80 + .49X$
2.13 Yes
2.14 Yes
2.15 $Y' = 3.62 + .483X$
2.16 (a) $SS_{reg} = 14.40$
(b) $SS_{res} = 25.60$
(c) .36
2.17 $c_{XY} = 0$
2.18 $r = 0$
2.19 Yes
2.20 $\Sigma(X - \bar{X})(Y - \bar{Y})$
$= \Sigma x(Y - \bar{Y})$
$= \Sigma xY - \bar{Y}\Sigma x$
$= \Sigma xY$ because $\bar{Y}\Sigma x = 0$

Chapter 3

3.1 No, because $s^2_{z'_1} = r^2$.
3.2 Only if $r_{12} = 1.0$, because $r_{z'_1 X_2}$ will always be equal to 1.0; z'_1 is just a linear transformation of X_2.
3.3 Yes, because z_1 is a linear transformation of X_2.
3.4 No, but see the answer to 3.5.
3.5 Yes. Try, for example, $r_{13} = 0$.
3.6 See text, Section 3.3
3.7 See text, Section 3.6.
3.8 In general, yes. But it is possible for $r_{12.3}$ to be larger than r_{12}. See the answer to 3.14.
3.9 See text, Section 3.5
3.10 The values of s_Y and s_X.
3.11 .36
3.12 Both will be equal to $1 - r^2$.
3.13 Both will be equal to r^2.
3.14 Yes. Try, for example, $r_{12} = .80$, $r_{13} = .60$, and $r_{23} = 0$.
3.15 The limits can be found by solving for the two roots of a quadratic equation and are given by

$$r_{23} = r_{12}r_{13} \pm \sqrt{1 - r_{12}^2}\sqrt{1 - r_{13}^2}$$

The limits for the values given are −.15 and .99.
3.16 $r_{1(2.3)} = .75$ and $r_{1(2.3)} = .80$

Chapter 4

4.1 (a) $R = .80$
(b) Yes
4.2 (a) $r^2_{Y(1.234)} = R^2_{Y.1234} - R^2_{Y.234}$
(b) $r^2_{Y1.234} = (R^2_{Y.1234} - R^2_{Y.234})/(1 - R^2_{Y.234})$
4.4 No

4.5 $R_{Y.123} = .748$

4.6 (a) $b_1 = 2.8$ and $b_2 = 1.2$
(b) Yes
(c) Yes
(d) $F = 48.17$
(e) $F = 48.17$
(f) .49
(g) .36
(h) .85

4.7 (a) $r_{Y1} = .50, r_{Y2} = .50,$ and $r_{12} = .50$
(b) $b_1 = 1.3333$ and $b_2 = .6667$
(c) $F = 4.35$
(d) $F = 2.13$
(e) $F = 2.13$
(f) Yes

Chapter 5

5.1 $\mathbf{I}^{-1} = \mathbf{I}$

5.2 No

5.3 $R_{2.13} = 1.0$

5.4 $\mathbf{R}^{-1} = \mathbf{I}$

5.5 $r_{Y(1.23)} = r_{Y1}$

5.6 $R^2_{Y.1234} = r^2_{Y1} + r^2_{Y2} + r^2_{Y3} + r^2_{Y4}$

5.7 $\hat{b}_i = r_{Yi}$

5.8 See text, Section 5.13,

5.9 See text, Section 5.13.

5.12 (a) $\Sigma Y' = \hat{b}_1\Sigma z_1 + \hat{b}_2\Sigma z_2 + \cdots + \hat{b}_k\Sigma z_k = 0$ because $\Sigma z_i = 0$ for all z_i.
(b) $s^2_{Y'} = (\hat{b}^2_1\Sigma z^2_1 + \hat{b}^2_2\Sigma z^2_2 + 2\hat{b}_1\hat{b}_2\Sigma z_1 z_2)/(n - 1)$
$= (.1245)^2 + (.3325)^2 + 2(.1245)(.3325)(.3461)$
$= .1547$
which is equal to R^2_{12}.
(c) In general, if all variables are in standardized form, then

$$s^2_{Y'} = \Sigma\hat{b}^2_i + 2\Sigma\hat{b}_i\hat{b}_j r_{ij} = R^2_{YY'}$$

Chapter 6

6.1 (a) $t = 6.00$
(b) $F = 36.00$
(c) $r^2 = .6667$
(d) $a = 4.0$
(e) $a = \bar{Y}_1$ and $b = \bar{Y}_1 - \bar{Y}_2$
(f) $SS_{reg} = 80$
(g) $F = 36.00$
(h) If $X = 0$, $Y' = \bar{Y}_2$.
If $X = 1$, $Y' = \bar{Y}_1$.
(i) $t^2 = 36.00$

6.2 (a) $t = 1.75$ and $F = 3.07$
(b) $r = .4847$ and $t = 1.75$
(c) $a = 3.0$, $b = 2.0$
(d) $t = 1.75$
(e) Yes

Chapter 7

7.1 (a) 3

(b) $a = \bar{Y}_4$, $b_1 = \bar{Y}_1 - \bar{Y}_4$, $b_2 = \bar{Y}_2 - \bar{Y}_4$, and $b_3 = \bar{Y} - \bar{Y}_4$

7.2 For any two X vectors, say X_1 and X_2, we have

$$\Sigma(X_1 - \bar{X}_1)(X_2 - \bar{X}_2) = 0 - \frac{n_i^2}{kn_i} = -n_i/k$$

and

$$\Sigma(X_1 - \bar{X}_1)^2 = n_i - \frac{n_i^2}{kn_i} = n_i - \frac{n_i}{k} = \Sigma(X_2 - \bar{X}_2)^2$$

Then

$$r_{12} = \frac{-n_i/k}{n_i - (n_i/k)} = -1/(k-1)$$

7.3 $R^2_{Y.123} = r^2_{Y1} + r^2_{Y2} + r^2_{Y3}$

7.4 (a) $SS_{tot} = 215$, $SS_W = 40$, $SS_T = 175$, and $F = 23.3333$

(b) $R^2_{Y.123} = .81395$

(c) $SS_T/SS_{tot} = .81395$

(d) $2151(1 - .81395) = 40$ and $(215)(.81395) = 175$

(e) $b_1 = 2$, $b_2 = 8$, and $b_3 = 4$

(f) $(2)(-7.5) + (8)(22.5) + (4)(2.5) = 175$

7.5 $R^2_{Y.123} = r^2_{Y1} + r^2_{Y2} + r^2_{Y3} = .81395$

7.7 (a) Yes. $\Sigma(X_i - \bar{X}_i)(X_j - \bar{X}_j) = 0$ for all pairs of X vectors.

(b) $R^2_{Y.123} = .70$

(c) $(\mu_1 + \mu_2) - (\mu_3 + \mu_4) = 0$

(d) $(\mu_1 + \mu_3) - (\mu_2 + \mu_4) = 0$

(e) $(\mu_1 + \mu_4) - (\mu_2 + \mu_3) = 0$

(f) $SS_T = 175$, $SS_W = 75$, and $F = 12.4444$

(g) $(-2.5)(-50) + (-1.5)(-30) + (.5)(10) = 175$

(h) $F = 26.6667$ for r_{Y1}; $F = 9.60$ for r_{Y2}; and $F = 1.0667$ for r_{Y3}

(i) $s_{b_i} = .4841$ and $t = -5.1642$ for b_1, $t = -3.0985$ for b_2, and $t = 1.0328$ for b_3

(j) Yes

Chapter 8

8.1 (a) $SS_T = 50$, $SS_W = 50$, and $F = 5$

(b) The sums of squares are 40.5, .36, 2.0, and 7.14, respectively.

(c) $b_1 = .90$, $b_2 = .07$, $b_3 = .20$, and $b_4 = -.14$

(d) $F = 16.2$ for the linear; $F = .144$ for the quadratic; $F = .80$ for the cubic; and $F = 2.86$ for the quartic

(e) $R^2_{Y.1234} = .810 + .007 + .040 + .143 = 1.00$

8.2 (a) $F = 5$
(b) Yes
(c) Yes
(d) Yes

Chapter 9

9.1 $F = MS_A/MS_W = 6.75$; $F = MS_B/MS_W = .75$; and $F = MS_{AB}/MS_W = .75$
9.2 (a) $r^2_{Y1} = .4154$; $r^2_{Y2} = .0462$; and $r^2_{Y3} = .0462$
(b) $R^2_{Y.123} = .5077$
(c) $SS_W/SS_{tot} = .4923$
(d) Yes
9.3 See text, Section 9.4.

Chapter 10

10.1 $b_1 = 1.1$
10.2 $b_2 = .50$
10.3 $a = 0$
10.4 $a = 0$
10.5 $F = MS_T/MS_{ST} = 6.3636$
10.6 (a) Yes
(b) $r^2_{Y1} = .390625$, $r^2_{Y2} = .046875$, and $r^2_{Y3} = .287500$
(c) $R^2_{Y.12} = .43750$
(d) $1 - R^2_{Y.123} = .27500$
(e) F = 6.3636
(f) $a = 0$
10.7 (a) $r_{Y1} = -.649519$, $r_{Y2} = -.216506$, and $r_{Y3} = .536190$
(b) $r_{12} = .50$; $R^2_{Y.12} = .43750$
(c) $b_1 = -1.667$ and $b_2 = .333$
(d) $b_3 = .333$
(e) $a = 0$
10.8 $\overline{r_{ij}s_is_j} = .6667$

Chapter 11

11.1 See text. Section 11.1.
11.2 See text. Section 11.3.
11.3

Source	d.f.
A	2
$S(A)$	12
B	2
$A \times B$	4
$S(A) \times B$	24

11.4 (a) It will require two vectors for A, two for B, four for the $A \times B$ interaction, and one for the sum vector.
(b) 14
(c) $SS_{S(A)} = SS_{tot}(r^2_{Y9} - R^2_{Y.12})$

11.5 (a) 12
(b) 20
(c) See text, Section 11.6.

11.7 For A_1 we have $\bar{c}_{ij} = 4.25$, and for A_2 we have $\bar{c}_{ij} = 4.75$. Then $\bar{c}_{ij}$, averaged over both levels of A, is equal to 4.5. Then we have $MS_{S(A)} - MS_{S(A)B} = 10.2 - 1.2 = 9.0$, which is equal to $k\bar{c}_{ij} = (2)(4.5)$.

11.8 (a) $MS_W = 91.2/16 = 5.7$
$MS_{S(A)B} = MS_W - \bar{c}_{ij} = 5.7 - 4.5 = 1.2$
$MS_{S(A)} = MS_W + (k - 1)\bar{c}_{ij} = 5.7 + (2 - 1)(4.5) = 10.2$
(b) $MS_{S(A)} - MS_{S(A)B} = k\bar{c}_{ij}$

Chapter 12

12.1 (a) $SS_T = 71.83333$, $SS_W = 36.0$
(b) $F = 7.1833$

Chapter 13

13.1 (a) $r^2_{Y1} = .6410$, $r^2_{Y2} = .2308$, and $r^2_{Y3} = .0256$
(b) $R^2_{Y.123} = .8974$

Chapter 14

14.1 (a) $1 - R^2_{Y.123}$
(b) $1 - R^2_{Y.12}$
(c) $r^2_{Y(3.12)}$
(d) $1 - R^2_{Y.1}$
(e) $r^2_{Y(2.1)}$

14.2 See text, Section 14.5.

14.3 See text, Section 14.6.

14.4 See text, Section 14.6.
(a) $SS_1 = 5.35$
(b) $SS_2 = 5.51$
(c) $b_1 = 1.35$, $b_2 = 1.49$, and $b_3 = 1.46$, $F = .135$
(d) $SS_4 = 264.41$
(e) $F = 258.43$

14.6 (a) $F = .136$
(b) $F = 258.47$

Chapter 15

15.1 (a) No
(b) Yes
(c) No
(d) $SS_A = SS_{tot}R^2_{Y.12}$, $SS_B = SS_{tot}(R^2_{Y.123} - R^2_{Y.12})$,
$SS_{AB} = SS_{tot}(R^2_{Y.12345} - R^2_{Y.123})$, $SS_W = SS_{tot}(1 - R^2_{Y.12345})$

15.2 (a) No
(b) Yes
(c) Yes
(d) $SS_A = SS_{tot}R^2_{Y.12}$, $SS_B = SS_{tot}(R^2_{Y.123} - R^2_{Y.12})$,
$SS_{AB} = SS_{tot}(R^2_{Y.12345} - R^2_{Y.123})$, $SS_W = SS_{tot}(1 - R^2_{Y.12345})$

15.3 (a) Yes
(b) Yes
(c) Yes
(d) SS_{ABC}

15.4 (a) $F = \dfrac{(R^2_{Y.12345} - R^2_{Y.123})/2}{(1 - R^2_{Y.12345})/(n - 5 - 1)}$
(b) See text, Section 15.5.

Appendix

TABLE I Table of t

The probabilities given by the column headings are for a one-sided test, assuming a null hypothesis to be true. For example, with 30 d.f., we have $P(t \geqslant 2.042) = .025$. For a two-sided test, we have $P(t \geqslant 2.042) + P(t \leqslant -2.042) = .025 + .025 = .05$.

d.f. \ P	.25	.10	.05	.025	.01	.005	.0025	.001
1	1.000	3.078	6.314	12.706	31.821	63.657	127.321	318.309
2	.816	1.886	2.920	4.303	6.965	9.925	14.089	22.327
3	.765	1.638	2.353	3.182	4.541	5.841	7.453	10.214
4	.741	1.533	2.132	2.776	3.747	4.604	5.598	7.173
5	.727	1.476	2.015	2.571	3.365	4.032	4.773	5.893
6	.718	1.440	1.943	2.447	3.143	3.707	4.317	5.208
7	.711	1.415	1.895	2.365	2.998	3.499	4.029	4.785
8	.706	1.397	1.860	2.306	2.896	3.355	3.833	4.501
9	.703	1.383	1.833	2.262	2.821	3.250	3.690	4.297
10	.700	1.372	1.812	2.228	2.764	3.169	3.581	4.144
11	.697	1.363	1.796	2.201	2.718	3.106	3.497	4.025
12	.695	1.356	1.782	2.179	2.681	3.055	3.428	3.930
13	.694	1.350	1.771	2.160	2.650	3.012	3.372	3.852
14	.692	1.345	1.761	2.145	2.624	2.977	3.326	3.787
15	.691	1.341	1.753	2.131	2.602	2.947	3.286	3.733
16	.690	1.337	1.746	2.120	2.583	2.921	3.252	3.686
17	.689	1.333	1.740	2.110	2.567	2.898	3.223	3.646
18	.688	1.330	1.734	2.101	2.552	2.878	3.197	3.610
19	.688	1.328	1.729	2.093	2.539	2.861	3.174	3.579
20	.687	1.325	1.725	2.086	2.528	2.845	3.153	3.552
21	.686	1.323	1.721	2.080	2.518	2.831	3.153	3.527
22	.686	1.321	1.717	2.074	2.508	2.819	3.119	3.505
23	.685	1.319	1.714	2.069	2.500	2.807	3.104	3.485
24	.685	1.318	1.711	2.064	2.492	2.797	3.090	3.467
25	.684	1.316	1.708	2.060	2.485	2.787	3.078	3.450
26	.684	1.315	1.706	2.056	2.479	2.779	3.067	3.435
27	.684	1.314	1.703	2.052	2.473	2.771	3.057	3.421
28	.683	1.313	1.701	2.048	2.467	2.763	3.047	3.408
29	.683	1.311	1.699	2.045	2.462	2.756	3.038	3.396
30	.683	1.310	1.697	2.042	2.457	2.750	3.030	3.385
35	.682	1.306	1.690	2.030	2.438	2.724	2.996	3.340
40	.681	1.303	1.684	2.021	2.423	2.704	2.971	3.307
45	.680	1.301	1.679	2.014	2.412	2.690	2.952	3.281
50	.679	1.299	1.676	2.009	2.403	2.678	2.937	3.261
55	.679	1.297	1.673	2.004	2.396	2.668	2.925	3.245
60	.679	1.296	1.671	2.000	2.390	2.660	2.915	3.232
70	.678	1.294	1.667	1.994	2.381	2.648	2.899	3.211
80	.678	1.292	1.664	1.990	2.374	2.639	2.887	3.195
90	.677	1.291	1.662	1.987	2.368	2.632	2.878	3.183
100	.677	1.290	1.660	1.984	2.364	2.626	2.871	3.174
200	.676	1.286	1.652	1.972	2.345	2.601	2.838	3.131
500	.675	1.283	1.648	1.965	2.334	2.586	2.820	3.107
1,000	.675	1.282	1.646	1.962	2.330	2.581	2.813	3.098
2,000	.675	1.282	1.645	1.961	2.328	2.578	2.810	3.094
10,000	.675	1.282	1.645	1.960	2.327	2.576	2.808	3.091
∞	.674	1.282	1.645	1.960	2.326	2.576	2.807	3.090

SOURCE: Reprinted from Enrico T. Federighi, Extended tables of the percentage points of student's t distribution. *Journal of the American Statistical Association,* 1959, *54,* 683–688, by permission.

TABLE I Table of t (*continued*)

d.f. \ P	.0005	.00025	.0001	.00005	.000025	.00001
1	636.619	1,273.239	3,183.099	6,366.198	12,732.395	31,830.989
2	31.598	44.705	70.700	99.992	141.416	223.603
3	12.924	16.326	22.204	28.000	35.298	47.928
4	8.610	10.306	13.034	15.544	18.522	23.332
5	6.869	7.976	9.678	11.178	12.893	15.547
6	5.959	6.788	8.025	9.082	10.261	12.032
7	5.408	6.082	7.063	7.885	8.782	10.103
8	5.041	5.618	6.442	7.120	7.851	8.907
9	4.781	5.291	6.010	6.594	7.215	8.102
10	4.587	5.049	5.694	6.211	6.757	7.527
11	4.437	4.863	5.453	5.921	6.412	7.098
12	4.318	4.716	5.263	5.694	6.143	6.756
13	4.221	4.597	5.111	5.513	5.928	6.501
14	4.140	4.499	4.985	5.363	5.753	6.287
15	4.073	4.417	4.880	5.239	5.607	6.109
16	4.015	4.346	4.791	5.134	5.484	5.960
17	3.965	4.286	4.714	5.044	5.379	5.832
18	3.922	4.233	4.648	4.966	5.288	5.722
19	3.883	4.187	4.590	4.897	5.209	5.627
20	3.850	4.146	4.539	4.837	5.139	5.543
21	3.819	4.110	4.493	4.784	5.077	5.469
22	3.792	4.077	4.452	4.736	5.022	5.402
23	3.768	4.048	4.415	4.693	4.972	5.343
24	3.745	4.021	4.382	4.654	4.927	5.290
25	3.725	3.997	4.352	4.619	4.887	5.241
26	3.707	3.974	4.324	4.587	4.850	5.197
27	3.690	3.954	4.299	4.558	4.816	5.157
28	3.674	3.935	4.275	4.530	4.784	5.120
29	3.659	3.918	4.254	4.506	4.756	5.086
30	3.646	3.902	4.234	4.482	4.729	5.054
35	3.591	3.836	4.153	4.389	4.622	4.927
40	3.551	3.788	4.094	4.321	4.544	4.835
45	3.520	3.752	4.049	4.269	4.485	4.766
50	3.496	3.723	4.014	4.228	4.438	4.711
55	3.476	3.700	3.986	4.196	4.401	4.667
60	3.460	3.681	3.962	4.169	4.370	4.631
70	3.435	3.651	3.926	4.127	4.323	4.576
80	3.416	3.629	3.899	4.096	4.288	4.535
90	3.402	3.612	3.878	4.072	4.261	4.503
100	3.390	3.598	3.862	4.053	4.240	4.478
200	3.340	3.539	3.789	3.970	4.146	4.369
500	3.310	3.504	3.747	3.922	4.091	4.306
1,000	3.300	3.492	3.733	3.906	4.073	4.285
2,000	3.295	3.486	3.726	3.898	4.064	4.275
10,000	3.292	3.482	3.720	3.892	4.058	4.267
∞	3.291	3.481	3.719	3.891	4.056	4.265

TABLE I Table of *t* (*continued*)

d.f. \ P	.000005	.0000025	.000001	.0000005	.00000025	.0000001
1	63,661.977	127,323.954	318,309.886	636,619.772	1,273,239.545	3,183,098.862
2	316.225	447.212	707.106	999.999	1,414.213	2,236.068
3	60.397	76.104	103.299	130.155	163.989	222.572
4	27.771	33.047	41.578	49.459	58.829	73.986
5	17.897	20.591	24.771	28.477	32.734	39.340
6	13.555	15.260	17.830	20.047	22.532	26.286
7	11.215	12.437	14.241	15.764	17.447	19.932
8	9.782	10.731	12.110	13.257	14.504	16.320
9	8.827	9.605	10.720	11.637	12.623	14.041
10	8.150	8.812	9.752	10.516	11.328	12.492
11	7.648	8.227	9.043	9.702	10.397	11.381
12	7.261	7.780	8.504	9.085	9.695	10.551
13	6.995	7.427	8.082	8.604	9.149	9.909
14	6.706	7.142	7.743	8.218	8.713	9.400
15	6.502	6.907	7.465	7.903	8.358	8.986
16	6.330	6.711	7.233	7.642	8.064	8.645
17	6.184	6.545	7.037	7.421	7.817	8.358
18	6.059	6.402	6.869	7.232	7.605	8.115
19	5.949	6.278	6.723	7.069	7.423	7.905
20	5.854	6.170	6.597	6.927	7.265	7.723
21	5.769	6.074	6.485	6.802	7.126	7.564
22	5.694	5.989	6.386	6.692	7.003	7.423
23	5.627	5.913	6.297	6.593	6.893	7.298
24	5.566	5.845	6.218	6.504	6.795	7.185
25	5.511	5.783	6.146	6.424	6.706	7.085
26	5.461	5.726	6.081	6.352	6.626	6.993
27	5.415	5.675	6.021	6.286	6.553	6.910
28	5.373	5.628	5.967	6.225	6.486	6.835
39	5.335	5.585	5.917	6.170	6.426	6.765
30	5.299	5.545	5.871	6.119	6.369	6.701
35	5.156	5.385	5.687	5.915	6.143	6.447
40	5.053	5.269	5.554	5.768	5.983	6.266
45	4.975	5.182	5.454	5.659	5.862	6.130
50	4.914	5.115	5.377	5.573	5.769	6.025
55	4.865	5.060	5.315	5.505	5.694	5.942
60	4.825	5.015	5.264	5.449	5.633	5.873
70	4.763	4.946	5.185	5.363	5.539	5.768
80	4.717	4.896	5.128	5.300	5.470	5.691
90	4.682	4.857	5.084	5.252	5.417	5.633
100	4.654	4.826	5.049	5.214	5.376	5.587
200	4.533	4.692	4.897	5.048	5.196	5.387
500	4.463	4.615	4.810	4.953	5.094	5.273
1,000	4.440	4.590	4.781	4.922	5.060	5.236
2,000	4.428	4.578	4.767	4.907	5.043	5.218
10,000	4.419	4.567	4.756	4.895	5.029	5.203
∞	4.417	4.565	4.753	4.892	5.026	5.199

TABLE II Values of F significant with $\alpha = .05$ and $\alpha = .01$

The values of F significant at the .05 (roman type) and .01 (**boldface type**) levels of significance with n_1 degrees of freedom for the numerator and n_2 degrees of freedom for the denominator of the F ratio.

	n_1 Degrees of freedom																							
n_2	1	2	3	4	5	6	7	8	9	10	11	12	14	16	20	24	30	40	50	75	100	200	500	∞
1	161	200	216	225	230	234	237	239	241	242	243	244	245	246	248	249	250	251	252	253	253	254	254	254
	4,052	**4,999**	**5,403**	**5,625**	**5,764**	**5,859**	**5,928**	**5,981**	**6,022**	**6,056**	**6,082**	**6,106**	**6,142**	**6,169**	**6,208**	**6,234**	**6,258**	**6,286**	**6,302**	**6,323**	**6,334**	**6,352**	**6,361**	**6,366**
2	18.51	19.00	19.16	19.25	19.30	19.33	19.36	19.37	19.38	19.39	19.40	19.41	19.42	19.43	19.44	19.45	19.46	19.47	19.47	19.48	19.49	19.49	19.50	19.50
	98.49	**99.00**	**99.17**	**99.25**	**99.30**	**99.33**	**99.34**	**99.36**	**99.38**	**99.40**	**99.41**	**99.42**	**99.43**	**99.44**	**99.45**	**99.46**	**99.47**	**99.48**	**99.48**	**99.49**	**99.49**	**99.49**	**99.50**	**99.50**
3	10.13	9.55	9.28	9.12	9.01	8.94	8.88	8.84	8.81	8.78	8.76	8.74	8.71	8.69	8.66	8.64	8.62	8.60	8.58	8.57	8.56	8.54	8.54	8.53
	34.12	**30.82**	**29.46**	**28.71**	**28.24**	**27.91**	**27.67**	**27.49**	**27.34**	**27.23**	**27.13**	**27.05**	**26.92**	**26.83**	**26.69**	**26.60**	**26.50**	**26.41**	**26.35**	**26.27**	**26.23**	**26.18**	**26.14**	**26.12**
4	7.71	6.94	6.59	6.39	6.26	6.16	6.09	6.04	6.00	5.96	5.93	5.91	5.87	5.84	5.80	5.77	5.74	5.71	5.70	5.68	5.66	5.65	5.64	5.63
	21.20	**18.00**	**16.69**	**15.98**	**15.52**	**15.21**	**14.98**	**14.80**	**14.66**	**14.54**	**14.45**	**14.37**	**14.24**	**14.15**	**14.02**	**13.93**	**13.83**	**13.74**	**13.69**	**13.61**	**13.57**	**13.52**	**13.48**	**13.46**
5	6.61	5.79	5.41	5.19	5.05	4.95	4.88	4.82	4.78	4.74	4.70	4.68	4.64	4.60	4.56	4.53	4.50	4.46	4.44	4.42	4.40	4.38	4.37	4.36
	16.26	**13.27**	**12.06**	**11.39**	**10.97**	**10.67**	**10.45**	**10.27**	**10.15**	**10.05**	**9.96**	**9.89**	**9.77**	**9.68**	**9.55**	**9.47**	**9.38**	**9.29**	**9.24**	**9.17**	**9.13**	**9.07**	**9.04**	**9.02**
6	5.99	5.14	4.76	4.53	4.39	4.28	4.21	4.15	4.10	4.06	4.03	4.00	3.96	3.92	3.87	3.84	3.81	3.77	3.75	3.72	3.71	3.69	3.68	3.67
	13.74	**10.92**	**9.78**	**9.15**	**8.75**	**8.47**	**8.26**	**8.10**	**7.98**	**7.87**	**7.79**	**7.72**	**7.60**	**7.52**	**7.39**	**7.31**	**7.23**	**7.14**	**7.09**	**7.02**	**6.99**	**6.94**	**6.90**	**6.88**
7	5.59	4.74	4.35	4.12	3.97	3.87	3.79	3.73	3.68	3.63	3.60	3.57	3.52	3.49	3.44	3.41	3.38	3.34	3.32	3.29	3.28	3.25	3.24	3.23
	12.25	**9.55**	**8.45**	**7.85**	**7.46**	**7.19**	**7.00**	**6.84**	**6.71**	**6.62**	**6.54**	**6.47**	**6.35**	**6.27**	**6.15**	**6.07**	**5.98**	**5.90**	**5.85**	**5.78**	**5.75**	**5.70**	**5.67**	**5.65**
8	5.32	4.46	4.07	3.84	3.69	3.58	3.50	3.44	3.39	3.34	3.31	3.28	3.23	3.20	3.15	3.12	3.08	3.05	3.03	3.00	2.98	2.96	2.94	2.93
	11.26	**8.65**	**7.59**	**7.01**	**6.63**	**6.37**	**6.19**	**6.03**	**5.91**	**5.82**	**5.74**	**5.67**	**5.56**	**5.48**	**5.36**	**5.28**	**5.20**	**5.11**	**5.06**	**5.00**	**4.96**	**4.91**	**4.88**	**4.86**
9	5.12	4.26	3.86	3.63	3.48	3.37	3.29	3.23	3.18	3.13	3.10	3.07	3.02	2.98	2.93	2.90	2.86	2.82	2.80	2.77	2.76	2.73	2.72	2.71
	10.56	**8.02**	**6.99**	**6.42**	**6.06**	**5.80**	**5.62**	**5.47**	**5.35**	**5.26**	**5.18**	**5.11**	**5.00**	**4.92**	**4.80**	**4.73**	**4.64**	**4.56**	**4.51**	**4.45**	**4.41**	**4.36**	**4.33**	**4.31**
10	4.96	4.10	3.71	3.48	3.33	3.22	3.14	3.07	3.02	2.97	2.94	2.91	2.86	2.82	2.77	2.74	2.70	2.67	2.64	2.61	2.59	2.56	2.55	2.54
	10.04	**7.56**	**6.55**	**5.99**	**5.64**	**5.39**	**5.21**	**5.06**	**4.95**	**4.85**	**4.78**	**4.71**	**4.60**	**4.52**	**4.41**	**4.33**	**4.25**	**4.17**	**4.12**	**4.05**	**4.01**	**3.96**	**3.93**	**3.91**
11	4.84	3.98	3.59	3.36	3.20	3.09	3.01	2.95	2.90	2.86	2.82	2.79	2.74	2.70	2.65	2.61	2.57	2.53	2.50	2.47	2.45	2.42	2.41	2.40
	9.65	**7.20**	**6.22**	**5.67**	**5.32**	**5.07**	**4.88**	**4.74**	**4.63**	**4.54**	**4.46**	**4.40**	**4.29**	**4.21**	**4.10**	**4.02**	**3.94**	**3.86**	**3.80**	**3.74**	**3.70**	**3.66**	**3.62**	**3.60**
12	4.75	3.88	3.49	3.26	3.11	3.00	2.92	2.85	2.80	2.76	2.72	2.69	2.64	2.60	2.54	2.50	2.46	2.42	2.40	2.36	2.35	2.32	2.31	2.30
	9.33	**6.93**	**5.95**	**5.41**	**5.06**	**4.82**	**4.65**	**4.50**	**4.39**	**4.30**	**4.22**	**4.16**	**4.05**	**3.98**	**3.86**	**3.78**	**3.70**	**3.61**	**3.56**	**3.49**	**3.46**	**3.41**	**3.38**	**3.36**
13	4.67	3.80	3.41	3.18	3.02	2.92	2.84	2.77	2.72	2.67	2.63	2.60	2.55	2.51	2.46	2.42	2.38	2.34	2.32	2.28	2.26	2.24	2.22	2.21
	9.07	**6.70**	**5.74**	**5.20**	**4.86**	**4.62**	**4.44**	**4.30**	**4.19**	**4.10**	**4.02**	**3.96**	**3.85**	**3.78**	**3.67**	**3.59**	**3.51**	**3.42**	**3.37**	**3.30**	**3.27**	**3.21**	**3.18**	**3.16**
14	4.60	3.74	3.34	3.11	2.96	2.85	2.77	2.70	2.65	2.60	2.56	2.53	2.48	2.44	2.39	2.35	2.31	2.27	2.24	2.21	2.19	2.16	2.14	2.13
	8.86	**6.51**	**5.56**	**5.03**	**4.69**	**4.46**	**4.28**	**4.14**	**4.03**	**3.94**	**3.86**	**3.80**	**3.70**	**3.62**	**3.51**	**3.43**	**3.34**	**3.26**	**3.21**	**3.14**	**3.11**	**3.06**	**3.02**	**3.00**
15	4.54	3.68	3.29	3.06	2.90	2.79	2.70	2.64	2.59	2.55	2.51	2.48	2.43	2.39	2.33	2.29	2.25	2.21	2.18	2.15	2.12	2.10	2.08	2.07
	8.68	**6.36**	**5.42**	**4.89**	**4.56**	**4.32**	**4.14**	**4.00**	**3.89**	**3.80**	**3.73**	**3.67**	**3.56**	**3.48**	**3.36**	**3.29**	**3.20**	**3.12**	**3.07**	**3.00**	**2.97**	**2.92**	**2.89**	**2.87**
16	4.49	3.63	3.24	3.01	2.85	2.74	2.66	2.59	2.54	2.49	2.45	2.42	2.37	2.33	2.28	2.24	2.20	2.16	2.13	2.09	2.07	2.04	2.02	2.01
	8.53	**6.23**	**5.29**	**4.77**	**4.44**	**4.20**	**4.03**	**3.89**	**3.78**	**3.69**	**3.61**	**3.55**	**3.45**	**3.37**	**3.25**	**3.18**	**3.10**	**3.01**	**2.96**	**2.89**	**2.86**	**2.80**	**2.77**	**2.75**

TABLE II Values of F significant with $\alpha = .05$ and $\alpha = .01$ (*continued*)

n_2	n_1 Degrees of freedom																							
	1	2	3	4	5	6	7	8	9	10	11	12	14	16	20	24	30	40	50	75	100	200	500	∞
17	4.45	3.59	3.20	2.96	2.81	2.70	2.62	2.55	2.50	2.45	2.41	2.38	2.33	2.29	2.23	2.19	2.15	2.11	2.08	2.04	2.02	1.99	1.97	1.96
	8.40	**6.11**	**5.18**	**4.67**	**4.34**	**4.10**	**3.93**	**3.79**	**3.68**	**3.59**	**3.52**	**3.45**	**3.35**	**3.27**	**3.16**	**3.08**	**3.00**	**2.92**	**2.86**	**2.79**	**2.76**	**2.70**	**2.67**	**2.65**
18	4.41	3.55	3.16	2.93	2.77	2.66	2.58	2.51	2.46	2.41	2.37	2.34	2.29	2.25	2.19	2.15	2.11	2.07	2.04	2.00	1.98	1.95	1.93	1.92
	8.28	**6.01**	**5.09**	**4.58**	**4.25**	**4.01**	**3.85**	**3.71**	**3.60**	**3.51**	**3.44**	**3.37**	**3.27**	**3.19**	**3.07**	**3.00**	**2.91**	**2.83**	**2.78**	**2.71**	**2.68**	**2.62**	**2.59**	**2.57**
19	4.38	3.52	3.13	2.90	2.74	2.63	2.55	2.48	2.43	2.38	2.34	2.31	2.26	2.21	2.15	2.11	2.07	2.02	2.00	1.96	1.94	1.91	1.90	1.88
	8.18	**5.93**	**5.01**	**4.50**	**4.17**	**3.94**	**3.77**	**3.63**	**3.52**	**3.43**	**3.36**	**3.30**	**3.19**	**3.12**	**3.00**	**2.92**	**2.84**	**2.76**	**2.70**	**2.63**	**2.60**	**2.54**	**2.51**	**2.49**
20	4.35	3.49	3.10	2.87	2.71	2.60	2.52	2.45	2.40	2.35	2.31	2.28	2.23	2.18	2.12	2.08	2.04	1.99	1.96	1.92	1.90	1.87	1.85	1.84
	8.10	**5.85**	**4.94**	**4.43**	**4.10**	**3.87**	**3.71**	**3.56**	**3.45**	**3.37**	**3.30**	**3.23**	**3.13**	**3.05**	**2.94**	**2.86**	**2.77**	**2.69**	**2.63**	**2.56**	**2.53**	**2.47**	**2.44**	**2.42**
21	4.32	3.47	3.07	2.84	2.68	2.57	2.49	2.42	2.37	2.32	2.28	2.25	2.20	2.15	2.09	2.05	2.00	1.96	1.93	1.89	1.87	1.84	1.82	1.81
	8.02	**5.78**	**4.87**	**4.37**	**4.04**	**3.81**	**3.65**	**3.51**	**3.40**	**3.31**	**3.24**	**3.17**	**3.07**	**2.99**	**2.88**	**2.80**	**2.72**	**2.63**	**2.58**	**2.51**	**2.47**	**2.42**	**2.38**	**2.36**
22	4.30	3.44	3.05	2.82	2.66	2.55	2.47	2.40	2.35	2.30	2.26	2.23	2.18	2.13	2.07	2.03	1.98	1.93	1.91	1.87	1.84	1.81	1.80	1.78
	7.94	**5.72**	**4.82**	**4.31**	**3.99**	**3.76**	**3.59**	**3.45**	**3.35**	**3.26**	**3.18**	**3.12**	**3.02**	**2.94**	**2.83**	**2.75**	**2.67**	**2.58**	**2.53**	**2.46**	**2.42**	**2.37**	**2.33**	**2.31**
23	4.28	3.42	3.03	2.80	2.64	2.53	2.45	2.38	2.32	2.28	2.24	2.20	2.14	2.10	2.04	2.00	1.96	1.91	1.88	1.84	1.82	1.79	1.77	1.76
	7.88	**5.66**	**4.76**	**4.26**	**3.94**	**3.71**	**3.54**	**3.41**	**3.30**	**3.21**	**3.14**	**3.07**	**2.97**	**2.89**	**2.78**	**2.70**	**2.62**	**2.53**	**2.48**	**2.41**	**2.37**	**2.32**	**2.28**	**2.26**
24	4.26	3.40	3.01	2.78	2.62	2.51	2.43	2.36	2.30	2.26	2.22	2.18	2.13	2.09	2.02	1.98	1.94	1.89	1.86	1.82	1.80	1.76	1.74	1.73
	7.82	**5.61**	**4.72**	**4.22**	**3.90**	**3.67**	**3.50**	**3.36**	**3.25**	**3.17**	**3.09**	**3.03**	**2.93**	**2.85**	**2.74**	**2.66**	**2.58**	**2.49**	**2.44**	**2.36**	**2.33**	**2.27**	**2.23**	**2.21**
25	4.24	3.38	2.99	2.76	2.60	2.49	2.41	2.34	2.28	2.24	2.20	2.16	2.11	2.06	2.00	1.96	1.92	1.87	1.84	1.80	1.77	1.74	1.72	1.71
	7.77	**5.57**	**4.68**	**4.18**	**3.86**	**3.63**	**3.46**	**3.32**	**3.21**	**3.13**	**3.05**	**2.99**	**2.89**	**2.81**	**2.70**	**2.62**	**2.54**	**2.45**	**2.40**	**2.32**	**2.29**	**2.23**	**2.19**	**2.17**
26	4.22	3.37	2.98	2.74	2.59	2.47	2.39	2.32	2.27	2.22	2.18	2.15	2.10	2.05	1.99	1.95	1.90	1.85	1.82	1.78	1.76	1.72	1.70	1.69
	7.72	**5.53**	**4.64**	**4.14**	**3.82**	**3.59**	**3.42**	**3.29**	**3.17**	**3.09**	**3.02**	**2.96**	**2.86**	**2.77**	**2.66**	**2.58**	**2.50**	**2.41**	**2.36**	**2.28**	**2.25**	**2.19**	**2.15**	**2.13**
27	4.21	3.35	2.96	2.73	2.57	2.46	2.37	2.30	2.25	2.20	2.16	2.13	2.08	2.03	1.97	1.93	1.88	1.84	1.80	1.76	1.74	1.71	1.68	1.67
	7.68	**5.49**	**4.60**	**4.11**	**3.79**	**3.56**	**3.39**	**3.26**	**3.14**	**3.06**	**2.98**	**2.93**	**2.83**	**2.74**	**2.63**	**2.55**	**2.47**	**2.38**	**2.33**	**2.25**	**2.21**	**2.16**	**2.12**	**2.10**
28	4.20	3.34	2.95	2.71	2.56	2.44	2.36	2.29	2.24	2.19	2.15	2.12	2.06	2.02	1.96	1.91	1.87	1.81	1.78	1.75	1.72	1.69	1.67	1.65
	7.64	**5.45**	**4.57**	**4.07**	**3.76**	**3.53**	**3.36**	**3.23**	**3.11**	**3.03**	**2.95**	**2.90**	**2.80**	**2.71**	**2.60**	**2.52**	**2.44**	**2.35**	**2.30**	**2.22**	**2.18**	**2.13**	**2.09**	**2.06**
29	4.18	3.33	2.93	2.70	2.54	2.43	2.35	2.28	2.22	2.18	2.14	2.10	2.05	2.00	1.94	1.90	1.85	1.80	1.77	1.73	1.71	1.68	1.65	1.64
	7.60	**5.42**	**4.54**	**4.04**	**3.73**	**3.50**	**3.33**	**3.20**	**3.08**	**3.00**	**2.92**	**2.87**	**2.77**	**2.68**	**2.57**	**2.49**	**2.41**	**2.32**	**2.27**	**2.19**	**2.15**	**2.10**	**2.06**	**2.03**
30	4.17	3.32	2.92	2.69	2.53	2.42	2.34	2.27	2.21	2.16	2.12	2.09	2.04	1.99	1.93	1.89	1.84	1.79	1.76	1.72	1.69	1.66	1.64	1.62
	7.56	**5.39**	**4.51**	**4.02**	**3.70**	**3.47**	**3.30**	**3.17**	**3.06**	**2.98**	**2.90**	**2.84**	**2.74**	**2.66**	**2.55**	**2.47**	**2.38**	**2.29**	**2.24**	**2.16**	**2.13**	**2.07**	**2.03**	**2.01**
32	4.15	3.30	2.90	2.67	2.51	2.40	2.32	2.25	2.19	2.14	2.10	2.07	2.02	1.97	1.91	1.86	1.82	1.76	1.74	1.69	1.67	1.64	1.61	1.59
	7.50	**5.34**	**4.46**	**3.97**	**3.66**	**3.42**	**3.25**	**3.12**	**3.01**	**2.94**	**2.86**	**2.80**	**2.70**	**2.62**	**2.51**	**2.42**	**2.34**	**2.25**	**2.20**	**2.12**	**2.08**	**2.02**	**1.98**	**1.96**
34	4.13	3.28	2.88	2.65	2.49	2.38	2.30	2.23	2.17	2.12	2.08	2.05	2.00	1.95	1.89	1.84	1.80	1.74	1.71	1.67	1.64	1.61	1.59	1.57
	7.44	**5.29**	**4.42**	**3.93**	**3.61**	**3.38**	**3.21**	**3.08**	**2.97**	**2.89**	**2.82**	**2.76**	**2.66**	**2.58**	**2.47**	**2.38**	**2.30**	**2.21**	**2.15**	**2.08**	**2.04**	**1.98**	**1.94**	**1.91**
36	4.11	3.26	2.86	2.63	2.48	2.36	2.28	2.21	2.15	2.10	2.06	2.03	1.98	1.93	1.87	1.82	1.78	1.72	1.69	1.65	1.62	1.59	1.56	1.55
	7.39	**5.25**	**4.38**	**3.89**	**3.58**	**3.35**	**3.18**	**3.04**	**2.94**	**2.86**	**2.78**	**2.72**	**2.62**	**2.54**	**2.43**	**2.35**	**2.26**	**2.17**	**2.12**	**2.04**	**2.00**	**1.94**	**1.90**	**1.87**
38	4.10	3.25	2.85	2.62	2.46	2.35	2.26	2.19	2.14	2.09	2.05	2.02	1.96	1.92	1.85	1.80	1.76	1.71	1.67	1.63	1.60	1.57	1.54	1.53
	7.35	**5.21**	**4.34**	**3.86**	**3.54**	**3.32**	**3.15**	**3.02**	**2.91**	**2.82**	**2.75**	**2.69**	**2.59**	**2.51**	**2.40**	**2.32**	**2.22**	**2.14**	**2.08**	**2.00**	**1.97**	**1.90**	**1.86**	**1.84**

Table II Values of F significant with $\alpha = .05$ and $\alpha = .01$ (*continued*)

	n_1 Degrees of freedom																						
n_2	1	2	3	4	5	6	7	8	9	10	11	12	14	16	20	24	30	40	50	75	100	200	500	∞
40	4.08	3.23	2.84	2.61	2.45	2.34	2.25	2.18	2.12	2.07	2.04	2.00	1.95	1.90	1.84	1.79	1.74	1.69	1.66	1.61	1.59	1.55	1.53	1.51
	7.31	**5.18**	**4.31**	**3.83**	**3.51**	**3.29**	**3.12**	**2.99**	**2.88**	**2.80**	**2.73**	**2.66**	**2.56**	**2.49**	**2.37**	**2.29**	**2.20**	**2.11**	**2.05**	**1.97**	**1.94**	**1.88**	**1.84**	**1.81**
42	4.07	3.22	2.83	2.59	2.44	2.32	2.24	2.17	2.11	2.06	2.02	1.99	1.94	1.89	1.82	1.78	1.73	1.68	1.64	1.60	1.57	1.54	1.51	1.49
	7.27	**5.15**	**4.29**	**3.80**	**3.49**	**3.26**	**3.10**	**2.96**	**2.86**	**2.77**	**2.70**	**2.64**	**2.54**	**2.46**	**2.35**	**2.26**	**2.17**	**2.08**	**2.02**	**1.94**	**1.91**	**1.85**	**1.80**	**1.78**
44	4.06	3.21	2.82	2.58	2.43	2.31	2.23	2.16	2.10	2.05	2.01	1.98	1.92	1.88	1.81	1.76	1.72	1.66	1.63	1.58	1.56	1.52	1.50	1.48
	7.24	**5.12**	**4.26**	**3.78**	**3.46**	**3.24**	**3.07**	**2.94**	**2.84**	**2.75**	**2.68**	**2.62**	**2.52**	**2.44**	**2.32**	**2.24**	**2.15**	**2.06**	**2.00**	**1.92**	**1.88**	**1.82**	**1.78**	**1.75**
46	4.05	3.20	2.81	2.57	2.42	2.30	2.22	2.14	2.09	2.04	2.00	1.97	1.91	1.87	1.80	1.75	1.71	1.65	1.62	1.57	1.54	1.51	1.48	1.46
	7.21	**5.10**	**4.24**	**3.76**	**3.44**	**3.22**	**3.05**	**2.92**	**2.82**	**2.73**	**2.66**	**2.60**	**2.50**	**2.42**	**2.30**	**2.22**	**2.13**	**2.04**	**1.98**	**1.90**	**1.86**	**1.80**	**1.76**	**1.72**
48	4.04	3.19	2.80	2.56	2.41	2.30	2.21	2.14	2.08	2.03	1.99	1.96	1.90	1.86	1.79	1.74	1.70	1.64	1.61	1.56	1.53	1.50	1.47	1.45
	7.19	**5.08**	**4.22**	**3.74**	**3.42**	**3.20**	**3.04**	**2.90**	**2.80**	**2.71**	**2.64**	**2.58**	**2.48**	**2.40**	**2.28**	**2.20**	**2.11**	**2.02**	**1.96**	**1.88**	**1.84**	**1.78**	**1.73**	**1.70**
50	4.03	3.18	2.79	2.56	2.40	2.29	2.20	2.13	2.07	2.02	1.98	1.95	1.90	1.85	1.78	1.74	1.69	1.63	1.60	1.55	1.52	1.48	1.46	1.44
	7.17	**5.06**	**4.20**	**3.72**	**3.41**	**3.18**	**3.02**	**2.88**	**2.78**	**2.70**	**2.62**	**2.56**	**2.46**	**2.39**	**2.26**	**2.18**	**2.10**	**2.00**	**1.94**	**1.86**	**1.82**	**1.76**	**1.71**	**1.68**
55	4.02	3.17	2.78	2.54	2.38	2.27	2.18	2.11	2.05	2.00	1.97	1.93	1.88	1.83	1.76	1.72	1.67	1.61	1.58	1.52	1.50	1.46	1.43	1.41
	7.12	**5.01**	**4.16**	**3.68**	**3.37**	**3.15**	**2.98**	**2.85**	**2.75**	**2.66**	**2.59**	**2.53**	**2.43**	**2.35**	**2.23**	**2.15**	**2.06**	**1.96**	**1.90**	**1.82**	**1.78**	**1.71**	**1.66**	**1.64**
60	4.00	3.15	2.76	2.52	2.37	2.25	2.17	2.10	2.04	1.99	1.95	1.92	1.86	1.81	1.75	1.70	1.65	1.59	1.56	1.50	1.48	1.44	1.41	1.39
	7.08	**4.98**	**4.13**	**3.65**	**3.34**	**3.12**	**2.95**	**2.82**	**2.72**	**2.63**	**2.56**	**2.50**	**2.40**	**2.32**	**2.20**	**2.12**	**2.03**	**1.93**	**1.87**	**1.79**	**1.74**	**1.68**	**1.63**	**1.60**
65	3.99	3.14	2.75	2.51	2.36	2.24	2.15	2.08	2.02	1.98	1.94	1.90	1.85	1.80	1.73	1.68	1.63	1.57	1.54	1.49	1.46	1.42	1.39	1.37
	7.04	**4.95**	**4.10**	**3.62**	**3.31**	**3.09**	**2.93**	**2.79**	**2.70**	**2.61**	**2.54**	**2.47**	**2.37**	**2.30**	**2.18**	**2.09**	**2.00**	**1.90**	**1.84**	**1.76**	**1.71**	**1.64**	**1.60**	**1.56**
70	3.98	3.13	2.74	2.50	2.35	2.23	2.14	2.07	2.01	1.97	1.93	1.89	1.84	1.79	1.72	1.67	1.62	1.56	1.53	1.47	1.45	1.40	1.37	1.35
	7.01	**4.92**	**4.08**	**3.60**	**3.29**	**3.07**	**2.91**	**2.77**	**2.67**	**2.59**	**2.51**	**2.45**	**2.35**	**2.28**	**2.15**	**2.07**	**1.98**	**1.88**	**1.82**	**1.74**	**1.69**	**1.62**	**1.56**	**1.53**
80	3.96	3.11	2.72	2.48	2.33	2.21	2.12	2.05	1.99	1.95	1.91	1.88	1.82	1.77	1.70	1.65	1.60	1.54	1.51	1.45	1.42	1.38	1.35	1.32
	6.96	**4.88**	**4.04**	**3.56**	**3.25**	**3.04**	**2.87**	**2.74**	**2.64**	**2.55**	**2.48**	**2.41**	**2.32**	**2.24**	**2.11**	**2.03**	**1.94**	**1.84**	**1.78**	**1.70**	**1.65**	**1.57**	**1.52**	**1.49**
100	3.94	3.09	2.70	2.46	2.30	2.19	2.10	2.03	1.97	1.92	1.88	1.85	1.79	1.75	1.68	1.63	1.57	1.51	1.48	1.42	1.39	1.34	1.30	1.28
	6.90	**4.82**	**3.98**	**3.51**	**3.20**	**2.99**	**2.82**	**2.69**	**2.59**	**2.51**	**2.43**	**2.36**	**2.26**	**2.19**	**2.06**	**1.98**	**1.89**	**1.79**	**1.73**	**1.64**	**1.59**	**1.51**	**1.46**	**1.43**
125	3.92	3.07	2.68	2.44	2.29	2.17	2.08	2.01	1.95	1.90	1.86	1.83	1.77	1.72	1.65	1.60	1.55	1.49	1.45	1.39	1.36	1.31	1.27	1.25
	6.84	**4.78**	**3.94**	**3.47**	**3.17**	**2.95**	**2.79**	**2.65**	**2.56**	**2.47**	**2.40**	**2.33**	**2.23**	**2.15**	**2.03**	**1.94**	**1.85**	**1.75**	**1.68**	**1.59**	**1.54**	**1.46**	**1.40**	**1.37**
150	3.91	3.06	2.67	2.43	2.27	2.16	2.07	2.00	1.94	1.89	1.85	1.82	1.76	1.71	1.64	1.59	1.54	1.47	1.44	1.37	1.34	1.29	1.25	1.22
	6.81	**4.75**	**3.91**	**3.44**	**3.14**	**2.92**	**2.76**	**2.62**	**2.53**	**2.44**	**2.37**	**2.30**	**2.20**	**2.12**	**2.00**	**1.91**	**1.83**	**1.72**	**1.66**	**1.56**	**1.51**	**1.43**	**1.37**	**1.33**
200	3.89	3.04	2.65	2.41	2.26	2.14	2.05	1.98	1.92	1.87	1.83	1.80	1.74	1.69	1.62	1.57	1.52	1.45	1.42	1.35	1.32	1.26	1.22	1.19
	6.74	**4.71**	**3.88**	**3.41**	**3.11**	**2.90**	**2.73**	**2.60**	**2.50**	**2.41**	**2.34**	**2.28**	**2.17**	**2.09**	**1.97**	**1.88**	**1.79**	**1.69**	**1.62**	**1.53**	**1.48**	**1.39**	**1.33**	**1.28**
400	3.86	3.02	2.62	2.39	2.23	2.12	2.03	1.96	1.90	1.85	1.81	1.78	1.72	1.67	1.60	1.54	1.49	1.42	1.38	1.32	1.28	1.22	1.16	1.13
	6.70	**4.66**	**3.83**	**3.36**	**3.06**	**2.85**	**2.69**	**2.55**	**2.46**	**2.37**	**2.29**	**2.23**	**2.12**	**2.04**	**1.92**	**1.84**	**1.74**	**1.64**	**1.57**	**1.47**	**1.42**	**1.32**	**1.24**	**1.19**
1000	3.85	3.00	2.61	2.38	2.22	2.10	2.02	1.95	1.89	1.84	1.80	1.76	1.70	1.65	1.58	1.53	1.47	1.41	1.36	1.30	1.26	1.19	1.13	1.08
	6.66	**4.62**	**3.80**	**3.34**	**3.04**	**2.82**	**2.66**	**2.53**	**2.43**	**2.34**	**2.26**	**2.20**	**2.09**	**2.01**	**1.89**	**1.81**	**1.71**	**1.61**	**1.54**	**1.44**	**1.38**	**1.28**	**1.19**	**1.11**
∞	3.84	2.99	2.60	2.37	2.21	2.09	2.01	1.94	1.88	1.83	1.79	1.75	1.69	1.64	1.57	1.52	1.46	1.40	1.35	1.28	1.24	1.17	1.17	1.00
	6.64	**4.60**	**3.78**	**3.32**	**3.02**	**2.80**	**2.64**	**2.51**	**2.41**	**2.32**	**2.24**	**2.18**	**2.07**	**1.99**	**1.87**	**1.79**	**1.69**	**1.59**	**1.52**	**1.41**	**1.36**	**1.25**	**1.15**	**1.00**

TABLE III Table of coefficients for orthogonal polynomials

k	Polynomial	Values of the Coefficients									
3	Linear	-1	0	1							
	Quadratic	1	-2	1							
4	Linear	-3	-1	1	3						
	Quadratic	1	-1	-1	1						
	Cubic	-1	3	-3	1						
5	Linear	-2	-1	0	1	2					
	Quadratic	2	-1	-2	-1	2					
	Cubic	-1	2	0	-2	1					
	Quartic	1	-4	6	-4	1					
6	Linear	-5	-3	-1	1	3	5				
	Quadratic	5	-1	-4	-4	-1	5				
	Cubic	-5	7	4	-4	-7	5				
	Quartic	1	-3	2	2	-3	1				
7	Linear	-3	-2	-1	0	1	2	3			
	Quadratic	5	0	-3	-4	-3	0	5			
	Cubic	-1	1	1	0	-1	-1	1			
	Quartic	3	-7	1	6	1	-7	3			
8	Linear	-7	-5	-3	-1	1	3	5	7		
	Quadratic	7	1	-3	-5	-5	-3	1	7		
	Cubic	-7	5	7	3	-3	-7	-5	7		
	Quartic	7	-13	-3	9	9	-3	-13	7		
9	Linear	-4	-3	-2	-1	0	1	2	3	4	
	Quadratic	28	7	-8	-17	-20	-17	-8	7	28	
	Cubic	-14	7	13	9	0	-9	-13	-7	14	
	Quartic	14	-21	-11	9	18	9	-11	-21	14	
10	Linear	-9	-7	-5	-3	-1	1	3	5	7	9
	Quadratic	6	2	-1	-3	-4	-4	-3	-1	2	6
	Cubic	-42	14	35	31	12	-12	-31	-35	-14	42
	Quartic	18	-22	-17	3	18	18	3	-17	-22	18

Index